Praise

"Mark Salvatore's Peace Corps memoir is a painstakingly intimate insight into a service most only know by name or from a bit of history. Salvatore offers the reader scrupulous detail and heartfelt colour with his story and, most importantly, truth without varnish, something in all too short a supply, these days."
ERIC KINKOPF, AWARD-WINNING JOURNALIST AND AUTHOR OF
SHOOTER AND *THE UNTIMELY DEATH OF MOLLY MACBETH*

"Rural Paraguay in all its simplicity, decades ago! Overcoming obstacles—like experiencing life on another planet. Enjoy this great read as Mark takes you on his incredible Peace Corps journey. The laughter, frustrations, cultural misunderstandings and language barriers all make for a story you won't forget."
MARCIA MAYER, RN, RETURNED PEACE CORPS VOLUNTEER,
RURAL NURSE PARAGUAY 1995-1997

About the Author

Mark Salvatore is a writer, a teacher and a former Peace Corps volunteer who served in Paraguay where he lived for several years. He is the author of a novel, *Labeled*. He lives with his wife in Deep South Texas where he researches and writes nonfiction.

SHADE
of the
PARAISO

Mark Salvatore

Vine Leaves Press
Melbourne, Vic, Australia

Shade of the Paraiso
Copyright © 2018 Mark Salvatore
All rights reserved.

Print Edition
ISBN:978-1-925417-66-1

Published by Vine Leaves Press 2018
Melbourne, Victoria, Australia

No parts of this publication may be reproduced, stored in a retrieval system, or transmitted in any form or by any means, electronic, mechanical, photocopying, recording, or otherwise, without the prior written permission of the copyright owner.

This book is sold subject to the condition that it shall not, by way of trade or otherwise, be lent, resold, hired out, or otherwise circulated without the publisher's prior consent in any form of binding or cover other than that in which it is published and without a similar condition including this condition being imposed on the subsequent purchaser. Under no circumstances may any part of this book be photocopied for resale.

Cover design by Jessica Bell
Interior design by Amie McCracken

 A catalogue record for this book is available from the National Library of Australia

For Noemi, Augusto and Elena

1

An open truck packed with armed troops in fatigues rolled toward downtown Asunción along the *Avenida de Aviadores del Chaco*. I refilled my glass with beer from a litre bottle and listened to a voice singing in Guaraní, an indigenous Paraguayan language, accompanied by guitars and a harp blaring through cracked speakers. Cigarette smoke and the scent of roasting chicken blended with exhaust fumes wafted through the patio of the restaurant. My shirt clung to my wet skin in the sauna-like February heat, even at half past nine at night. Busses raced past, honking, weaving through traffic, and leaving plumes of black diesel smoke in their wakes. Soon a tank rumbled toward downtown. A soldier sat upon the turret with an automatic rifle on his lap.

I sat in a restaurant at a table strewn with beer and soda bottles and glasses and plates, with several men and women from my Peace Corps group and one of our trainers. Another truck loaded with soldiers passed followed by a tank.

We had arrived in Paraguay that same day, Thursday, at 2:30 a.m., February 2, 1989, as a group of thirty-five aspiring Peace Corps volunteers. The troop movement appeared casual and routine, and, I thought, probably common.

"Something bad is happening," the proprietor said. "You have to leave. I'm closing."

We drained our beers. John, the trainer, paid the tab, and we stepped out from beneath the roof of the dining area and through the gate onto the sidewalk as the restaurant owner slammed shut a rusting iron gate and slid a deadbolt into place. We took a few steps toward Ykua Satí, a retreat Peace Corps Paraguay used for the acclimation of new groups of volunteers before sending them to the training centre in Areguá, a town on the *Lago Ypacaraí* (Lake Ypacaraí) about twenty kilometres east of Asunción, the capital city and Paraguay's most populous, where a half million of the country's five million residents lived.

Traffic sped by travelling away from downtown. The popping of automatic weapons and blasts from a tank snapped us to attention.

There had been some dissent within the Colorado Party, the long-standing ruling party in Paraguay, and the military convoy was probably a show of force meant to intimidate another political faction, John explained.

We left the main avenue and straggled back up a cobblestone road to Ykua Satí. We passed three women standing and talking in front of a house as we listened to tanks rumbling toward the downtown area. The women said they didn't know what was happening, but it wasn't good and that we shouldn't be out on the street. We continued up the hill to the retreat where we found the rest of our group and a couple of trainers, all sitting in a circle and listening to a radio beneath the canopy of a massive mango tree. The trainers, all former Peace Corps volunteers, translated the Spanish radio broadcast.

"It's just a power play," one said. "Major General Andrés Rodríguez wants to take power from President Stroessner." Stroessner had held power for thirty-four years since he staged his own military coup on May 4, 1954.

"This has been building for some years," another said. "You know, after Pope John Paul II visited here in '87 things began to change. Even before that, Stroessner's Colorado Party split."

"I thought that Stroessner closed some *Casas de Cambios* that Rodríguez owned and that's why Rodríguez brought out the troops," a third added, referring to currency exchange houses. Some reports alleged that Stroessner allowed favoured party members, such as General Rodríguez, to operate lucrative businesses. Favoured, loyal party members ran most of the profitable businesses in the country, legal or not.

A military jet swooped low and banked over our mango tree. "Did you see that?" someone yelled.

"See it? I waved to the pilot!" A woman said.

We stared at the jet. It flew low, fast, and loud. One of the trainers walked off with a walkie-talkie. Automatic weapons and tanks blasted the humid night awake.

"They're saying that people should stay in their houses," a trainer said, gesturing toward the radio. Stations broadcast the same message: Everything is safe, calm and tranquil. Please stay in your homes. The troops entering the city are on manoeuvres.

They had taken ANTELCO (then the national telephone company), the newspapers, and the radio and television stations, we were told. Stroessner had no control over the media.

Another trainer walked back into our circle beneath the mango with his walkie-talkie. "We have orders to stay and wait," he said. "If things get too bad, the embassy will airlift us out."

Radio stations announced that two divisions loyal to Stroessner went over to General Rodríguez and that the shooting had stopped a half hour ago, but stay in your homes. And the firing continued.

Another broadcast said that the main police barracks in Asunción had been shelled and burnt. An artillery unit from another *Departamento* arrived and fired on the guards near the Presidential Palace.

I looked at the faces near me. Each person sat immersed in thoughts unspoken. I hoped we wouldn't have to leave, and I wondered what people heard about this in the United States, if anything.

At around 3:00 a.m. we picked up the BBC on a shortwave radio. A reporter confirmed that there was heavy fighting in downtown Asunción and that the main police barracks were aflame. Apparently Stroessner tried to force General Rodríguez to retire ten days earlier and the coup had been building since then, someone speculated. We sat listening to gunfire and to the radio. I wondered if we would stay.

***

The Peace Corps application process from the first inquiry to an invitation to serve had taken nine months, and I'd left the United States three months after the invitation. I had quit my job and had given away most of what I owned. I sold my car for $300 and put the money in a bank. I left Albuquerque, New Mexico, for Miami, Florida, with the airline ticket Peace Corps provided and with $150 dollars of my own. I carried a full backpack and a hand bag. I arrived in Miami and met the other members of my group and the trainers and Peace Corps representatives. I was an American going overseas with little actual understanding of what my Peace Corps commitment meant. I thought I knew, since I'd spoken with former volunteers and had read books and articles. I had imagined Peace Corps dropping me into an exotic place where I would adapt to the local culture and help people improve their lives. My greatest challenge would be to help others lead healthier and more efficient lives without altering their culture. I hadn't considered that military violence in the host country might be part of my experience. Understanding another culture and its history would maybe become my biggest task as a volunteer.

We thought of ourselves as volunteers, yet Peace Corps Paraguay considered us *aspirantes*, meaning that we aspired to be volunteers. A few months of training would determine those of us who would later be known as volunteers.

Excitement, anticipation and, in my case, uncertainty, hung over the group like an aura. We talked, exchanged ideas, walked, danced, drank, and saw what we could of Miami before our plane left. John, a former volunteer and trainer had us tie bright orange yarn to all our bags so we could easily identify them at airports.

We started our training in Miami where we spent two days at a hotel. The trainers led some ice-breaking activities and we attempted to graphically identify ourselves on sheets of paper. A glance about and I saw that I was older than most and younger than some. Most were recently out of college. Some had retired. Our group had a few couples, some older singles. A common ideal—to make a difference—bound us.

We checked out of our rooms at the hotel and headed to the airport for our flight to Paraguay with stops in Panama, Peru, and Bolivia. I had nothing to return to in Albuquerque that would not require starting all over again. I wanted to stay in Paraguay even though I'd yet to arrive.

***

The sound of weapons firing and the blasts of tanks reached us until almost 4:00 a.m.

"Rodríguez controls the *Departamento de Caaguazú*," Charlie, a trainer, translated from the radio. "The *Departamento de Paraguarí*, too." Later, radio broadcasts announced the fall of all nineteen *Departamentos* in the *Republica Del Paraguay*. By 4:00 General Rodríguez had announced his victory on the radio. He said that power fell his way for God and for the people of Paraguay. He said that he promised a general election soon for a democratic Paraguay.

His thirty-five years of rule over, former Dictator Alfredo Stroessner waited for a country to take him as an exile. Brazil accepted him, and on February 5 he departed for Brasilia, where he would spend the rest of his life.

Leaders of Stroessner's regime had been arrested or had fled.

Quiet returned to Asunción. We all retired to our bunks and the drone of mosquitoes as we tried to sleep in the clammy air.

***

We had been assigned rooms at Ykua Satí, the men in one building and the women in another. A centre hall down the length of the plastered brick rectangular building separated twelve rooms, which measured about three by four metres, each furnished with bunk beds and two wardrobes. Ceiling fans spun sluggishly hung from high rafters. A light sultry breeze crossed the rooms between the open windows and the corridor. Water dripped from all the faucets in the bathrooms at the end of the building.

We arose Friday with the sun and the sounds of chickens cackling and pecking and of women sweeping and busses racing on the *Avenida*. Within an hour we gathered in the dining hall and ate golf ball-size hard rolls, cut and spread with butter and *dulce de guayaba* (guava jam) and coffee.

I sat beneath the mango tree after breakfast and listened to a few other volunteers.

"I'm a little worried about what my parents are hearing on the news," one said.

"My mum's probably already called the Peace Corps office in Washington," said another.

"I just don't feel good about being so out of touch, you know? And we're still in the city."

"I really want to talk to my boyfriend and it's only been a couple of days. We still have two years to go."

"Yeah, well, this will take getting used to, but I'm looking forward to the adventure even if my folks think I'm in danger here. I mean, I knew I'd be breaking ties. I volunteered to do something different."

"And I don't care one way or another as long as we don't get sent home."

***

Later in the morning we gathered in the dining hall to listen to Edgardo, director of the training centre.

"One day later and you would have missed the change of government," said the Chilean and former priest. "But that makes an introduction to recent events in Paraguay more appropriate."

"First, we—I and the trainers here— are contractors and we operate out of the Peace Corps training centre in Areguá, a town nearby. The Center for Human Potential, or CHP, our company, has done training for Peace Corps since 1978. Every few years we move the centre to ensure that any one town doesn't become too saturated by our presence. You're the group Areguá-1, since we just moved to that town. You'll be there soon."

"And what happened overnight was not unexpected," Edgardo explained. "Stroessner's party, the Colorado Party, split several years ago. Stroessner lost the support of many within his party, including General Rodríguez, the leader of the Army's 1st division."

"Rodríguez saw some of his coronels and some generals abruptly retire and then Stroessner asked him to retire. He refused and planned a coup. There's more to it than that, of course, but to be brief, I'll leave it there."

"Stroessner had been warned of a *coup d'état* but underestimated the threat," Edgardo said.

"Most of the fighting last night occurred near the U.S. Embassy and the Presidential Palace. Rodríguez claimed that no more than fifty died during the fighting. Radio Caritas reported that as many as 200 had died. Observers reported 300 and some said upwards of 1,000."

"The United States recognized the presidency of General Rodríguez. Everything appears to be normal now except for the soldiers with automatic weapons guarding the areas where the fighting occurred," Edgardo said.

***

Saturday night we had a party at Ykua Satí. Peace Corps provided food and drinks and a four-piece band with a Paraguayan harp (Paraguay produces some of the best harps in the world), a concertina, and two guitars. I waltzed and polkaed with a few of the women. Those of us on the dance floor formed a circle and then a line and danced around the room picking up people as we went until everyone in the group danced.

We packed in the morning to go to the training centre in Areguá but got off to a late start. Ten members of the group went out after the party and didn't return until 9:00 a.m. While we waited in the heavy moist air, breathing the aroma of blossoming plants and rotting mangoes, we loaded our bags on a waiting bus. When the ten returned and gathered their belongings, we boarded a second bus and rumbled toward downtown Asunción.

We drove directly to *Avenida Mariscal López* and saw the Peace Corps office and the American Embassy and plastered walls pockmarked from machine gun fire. Buildings stood with windows shot out. Piles of bricks from collapsed walls exposed destroyed homes. Soldiers with automatic weapons guarded barricaded streets at the area of the heaviest fighting. We passed the ANTELCO building downtown with its windows blown out and its telecommunications equipment destroyed. Bricklayers rebuilt walls that had been shelled. Otherwise, business downtown went on as usual.

The wide, green *Río Paraguay* (the Paraguay River) flowed past the *Palacio de López*, the presidential palace which was built and named by a former president. Other government buildings, docks and the customs house lined the high bank of the river. Just above the water line hundreds of shacks constructed from scrap lumber, plastic, and corrugated metal hugged the bank.

We left Asunción and drove about twenty kilometres east to Areguá, a town on Ypacaraí Lake—the largest lake in Paraguay—where we would stay with host families while in training. The two-lane road wound through green hills with tall, thin coconut trees, broad, leafy deciduous trees, and with bony Brahmas grazing in fields. As we dropped into Areguá we saw the long lake glistening in the sun. We passed colonial style homes with high tiled roofs and tall windows and broad verandahs. Ceramic shops and plant nurseries lined the street. Vendors sold fruit and vegetables from beneath canvas roofs on a corner. We turned onto a cobblestone, double *avenida* and drove toward the lake. Behind us the *avenida* led to Areguá's church, *Nuestra Señora de la Candelaria* (built in 1862) at the top of a hill overlooking the lake. We crossed railroad tracks and turned into the CHP training centre.

The bus with our luggage had already arrived and unloaded. We stepped down from the bus and walked to the area where we were to meet our families. They waited in the shade of trees and as we approached they applauded.

A woman called our names, and, one by one, we stepped forward.

2

Ña Fidencia, my Peace Corps training host, a short, stocky brown woman with long, grey braided hair and high cheek bones with skin stretched like crumpled leather, smiled and hugged me. The seventy-one-year-old lifted my backpack, but I wrestled it back from her and followed her as she padded along in *zapatillas* (flip flops) on a path and across the railroad tracks and up an uneven cobblestone road to her house, my home for the next twelve weeks.

We entered through an *almacen* (store), barely wide enough for a person to stand while making a purchase. A massive, stained hardwood bar separated the buyer from the merchant and the merchandise: hard rolls, flour, sugar, canned goods, wine, cigars and cigarettes, eggs, and in a refrigerator, beer, soda, and meat. Out the back door of the *almacen* we stepped into the kitchen, open on two sides. A sink and cabinets stood against one wall and a wood-burning stove formed a half-wall on one side. A blue and green parrot hung upside down from an iron ring strung from the beam of an open side of the kitchen. A path led to the house through guava and citrus trees and a grape arbour.

A verandah with a brick floor ran the width of the house. The whitewashed brick house had two spacious rooms with brick floors and high walls and door-size windows. The exterior walls stood twice my height. The tall wooden double doors in front were open and a door in the back was open. Outside the back door were two bedrooms, one on either side of the door, and a verandah along an addition to the house with a bathroom and a dining room. *Ña* Fidencia showed me my room, on the left beside the bathroom. A bed with a mosquito net and a *ropero* (wardrobe) filled the room except for a space to walk between the furnishings. A ceiling fan hung from a rafter.

No one in the house spoke English and I barely spoke Spanish. I met *Don* Pedro, *Ña* Fidencia's seventy-six-year-old husband, Juan Manuel, a grandson, and Silvestre, an adult son, all residents of the house. They left me to my room to unpack.

A bit after noon Juan Manuel called me to lunch. *Ña* Fidencia carried the food from the kitchen across the yard to the dining room. We ate rice with vegetables and beef chunks and *mandioca* (a fibrous root and rather tasteless except for a bland, nut-like essence. It's also known as yucca or cassava and is a staple in Paraguay). We also ate *dulce de guayaba* (guava jam) with cheese for dessert. A breeze blew through two opposite windows. From my seat at the table I saw in the yard two cows, a pig, chickens, geese, and a cage with a dozen guinea pigs. We communicated during the meal with the few words in Spanish I knew and through gestures.

I thought I asked *Ña* Fidencia in Spanish if the guinea pigs were food, because I had read that they are eaten in some countries. What I actually asked was if the guinea pigs ate. *Ña* Fidencia said that, yes, they ate a lot. I thought she meant that they, the family, ate them often. I counted the guinea pigs for a few days until I found out where my poor Spanish took me.

After lunch I paid *Ña* Fidencia 3,300 guaraníes (about thirty-three dollars at that time) for a ten-day stay. She said she'd wash my clothes and I offered her more guaraníes but she wouldn't take them until I insisted. I later learnt that the amount I paid per day included meals, a private bedroom with a fan, and the hand washing and ironing of my clothes, especially underclothes to kill a parasite known as *pique*. This parasite would burrow into a soft spot, usually on a toe, and lay its eggs, which resulted in a painful sore.

Ña Fidencia hung a new mosquito net above my bed and her son installed a light. Training would begin in the morning.

***

At 6:00 the next morning, the pig snorted and rooted in its pen. *Ña* Fidencia passed my window with a pan of feed on her shoulder.

"*Buen día*, Marcos," she said as she passed.

I'd been awake for about fifteen minutes when I heard her pounding meat. I knew she was up until at least midnight. I showered and then hauled a bucket of water from the well to use to flush the toilet. I walked to the kitchen and *Ña* Fidencia had small rolls, hot *yerba mate* (a tea-like beverage and a mild stimulant like coffee) and a banana waiting for me. I watched *Ña* Fidencia milk a cow while I ate. Pancho, the parrot, squawked at me from his iron ring.

I prepared for my first day of twelve training weeks. I held my notebook and pen and I thanked *Ña* Fidencia for breakfast. She left her cow, wiped her hands on her apron, and walked me to the door, telling me to return for lunch.

I stepped into the clammy morning heat and walked on uneven cobblestones in the shade of mango trees to the end of the road where I crossed a mossy creek on a plank foot bridge. I stepped over railroad tracks and took a path through tall weeds and across a marsh with a septic odour where frogs sounded like crying babies. Just beyond the marsh I reached the back entrance of the training centre grounds, six minutes from home. I entered the building, got my schedule from the office, and found my class.

We had classes from 7:45 a.m. to 5:15 p.m., excepting an hour and a half for lunch, during which time I returned home to eat. I started in a Spanish class of four. The teachers didn't, or wouldn't, speak English. They immersed us in Spanish and at home I had to speak Spanish, too. We also had daily lessons in Guaraní, and the first language of many Paraguayans. We learnt to communicate quickly.

In the afternoons, the Paraguayan and American trainers taught us Paraguayan customs or we had technical training according to our assignments. In my case, the training involved environmental sanitation. We built hand pumps from plastic piping and other easily acquired materials, we cultivated vegetable gardens, and we constructed sanitary latrines and a *fogón* (a wood-burning cooking stove). We learnt how to protect water sources, how to teach people about environmental sanitation, and how they might achieve a sanitary environment using appropriate technology, that is, with the materials at hand.

In teams, we started gardens at the training centre one week after our arrival. We gathered and hauled manure in a wheel barrow to our garden site and mixed it with the existing topsoil. We built *tablones*, or one by two metre raised beds, and planted spinach, onions, lettuce, and other vegetable seeds. We built a cover for our garden with bamboo and banana leaves to guard it from the direct sunlight. We took turns caring for our *tablones* and I had the first shift for my team. We then built a *losa*, a concrete slab with a hole in the middle for the floor of a latrine.

I wanted to assimilate into the community and the culture, but many of the group members usually got together after training. I attempted to stay to myself while longing for company with my fellow Americans.

Within a week we had language evaluations and interviews about our Paraguayan home life, our backgrounds, our motives, and about any changes we would like to make. The evaluations included a home visit during which my family counsellor from CHP interviewed *Ña* Fidencia and asked her questions about my behaviour and how I was progressing with the language and culture.

At night during the first couple of weeks I attended some local parties and met some of the local dancers who connected with a folk dance team. I had been an avid folk dancer in Albuquerque. They invited me to dance with them. I wanted to learn Paraguayan dances and they wanted to learn American dances, such as the swing.

***

We lost our first group member on Monday, February 20, just eighteen days after our arrival in Asunción. She said the Peace Corps wasn't right for her, she'd been depressed, she wasn't interested in learning the language, she was afraid of bugs, and she thought the Paraguayan people would be different. She missed her boyfriend and returned to her home in the Pacific Northwest.

Meanwhile, the lake reflected the full moon.

The next morning I awoke to solid sheets of grey rain and rolls of thunder that shook the house. Only the cracks of lightning lit the morning. Water cascaded from the roof leaving the ground beneath water. The road ran like a river. Trees bent to the wind. Only the clatter of rain remained constant. Most of us arrived at the training centre soaked. Even one person who was well prepared for the rain was drenched by a bus that sprayed him with the water it sped through. Rain fell throughout the day.

After school, I sat on the verandah with *Don* Pedro and he asked me if it rained like this where I came from, about my father, my car, and he wondered if there were gas stoves in New Mexico. He wanted to know if I liked Paraguay and he told me about Ñandutí, a Paraguayan embroidered lace. Ñandutí means, loosely, "spider web" in Guaraní. This would be our first conversation, yet I struggled for words. I hadn't learnt enough.

The next day, after almost three weeks as a guest in his house, I watched *Don* Pedro fall sick. I asked *Ña* Fidencia how he fared and she assured me that his health was fine. I watched the doctor leave and the family members arrive. *Ña* Fidencia invited me into *Don* Pedro's room. I entered.

Candles surrounded his bed as well as a statue of the *Virgin de Caacupé*. Family members sat around the bed conversing. A daughter-in-law wiped a cold rag on *Don* Pedro's forehead. He shook from fever, but he talked with his family. Everyone spoke in Guaraní and I suspected I sat at a death bed.

Don Pedro's dengue, a disease transmitted by an infected mosquito, passed and he resumed his place on the verandah, sitting in his shorts, tee-shirt, and *zapatillas*, watching people and cows pass along the cobblestone street.

The next weekend, a month into training, we would each visit another volunteer's site. I received a hand-sketched map and crude directions in Spanish. The site visits would take five days.

3

I left on Friday, March 5, by bus and travelled to the terminal in Asunción where busses spewing grey smoke careened through traffic, jerking to stops here and there to disgorge passengers and to pick up more. Charcoal smoke and the scent of cooking sausages and beef strips from sidewalk grills wafted over the area. Drivers in cars and trucks used their horns as often as their brakes. Street vendors accosted passersby. Police directing traffic blew whistles. Men swept crumpled cigarette and candy packs and dirty napkins from the gutters. Trees in the median of the street cast shadows on cobblestones and rutted asphalt outside of the long, two-storey bus terminal.

There, for 900 guaraníes, I caught a full bus going to La Colmena. I asked the driver to stop at *Tu Yú Jhú* (black mud in Guaraní), some kilometres before La Colmena. I stepped off the bus, crossed the highway, and started along a rutted dirt road toward Potrero Arce, eight kilometres distant.

***

As I walked, a man astride a horse approached. He wore a wide-brimmed floppy hat and muddy boots without laces. He reigned in his horse beside me and showed a few silver teeth as he smiled.

"Where are you going?" he asked me in Guaraní.

"Potrero Arce," I said, hoping I'd understood him.

"I'm going there, too," he said. "What are you going to Potrero for?"

"Is this the way," I said, pointing to the red, sandy road ahead. "I'm going to visit Juan, the Peace Corps volunteer. Do you know him?"

"Tall Juan. Plays guitar. Everyone knows Juan. Give me your backpack."

"My backpack?"

"I'll carry it for you. I'll take you to Juan. Are you related to Juan?"

"No, I'm in Peace Corps and want to visit him," I said, handing him my backpack. "Thanks for the help."

"*Ja ha*," he said. "Let's go." He tied my pack to the saddle and nudged his horse to begin walking. I fell in behind and followed, my feet sinking into the loose wet sand. We climbed steadily and in silence on the road cut through trees and fields of cotton plants.

We stopped at Peña, halfway to Potrero Arce, where a stocky woman with grey curls hanging from a torn, thin bandana tied on her head sat in front of her painted, straw-roofed, brick house washing clothes by hand. She grinned and nodded in greeting, her dark eyes shining from furrowed brown skin. She stood and brought out chairs for us and she prepared *tereré*, a drink of *yerba mate* in a *guampa*, a receptacle usually made from a cow's horn, into which loose *yerba mate* is packed and cold water poured over it. The liquid is sucked through a *bombilla*, a perforated metal straw. Once a person has sucked up the contents, the server refills the *guampa* with water and passes it to the next person, and so on. Once someone drinks, the youngest present or any woman refills the *guampa* and serves it. As we drank *tereré*, our host busied herself with other tasks. She would accept no money and we continued on our way.

We soon encountered a drunken man urinating on his pant leg and yelling. The road became mushy grass and we crossed several small clear creeks. At one point, I took off my shoes to wade 300 metres through knee-deep water and mud, an action against our training, because going barefoot invited hookworms. But I didn't want to get my shoes wet.

Late that afternoon we reached Potrero Arce. I met Juan (John), the Peace Corps volunteer. He had built a one-room house in the centre of the settlement, using the local style of brick with a straw roof. I washed up at his well and he led me to the *almacen* where I would spend my first night.

The locals gathered at the *almacen* in Potrero Arce. Everyone bought rice, flour, soda, beer and canned goods from there, and everyone sold their cotton there. Each Saturday morning the owner slaughtered a cow and sold the meat. In the evenings men met there to drink and talk.

Juan left me at the *almacen* at dusk. The *señora* served me rice with meat and *mandioca* and a cup of milk. As I ate, a group of men sat at another table in front of the *almacen* drinking and playing cards. They didn't speak to me, but they stared at me. I didn't speak to them, either, because of my lack of Guaraní.

I went to sleep in one of the two rooms of the house. Their fourteen-year-old son slept in the same room.

***

Before dawn I heard voices outside. I arose and saw men around a slaughtered cow. They cut and hung slabs of meat from hooks in the breezeway between the *almacen* and the rooms. A line of people waited to purchase meat as the vendors tared their scales. Tethered horses waited and grazed.

Representatives from most of the families who occupied the forty-five houses in this cotton farming community waited to buy meat.

I met a man and a woman who were waiting. The man's hair had bits of straw in it and it hung over his collar. He wore a wide cloth belt around his waist with a short machete hanging from it. The woman stood barefoot in a tattered paisley dress and a stained straw hat.

"Arnolfo," the man said, holding out his hand.

"Marco," I said, shaking his hand.

"And she's Aubrelia," he said, gesturing toward the woman. She smiled and shook my hand.

"You're Juan's friend?" Arnolfo said.

"Yes, I'm here visiting him. And you? Are you from here? Potrero Arce?"

"We were born here maybe forty years ago," Arnolfo said, glancing at Aubrelia.

"And do you have children?" I asked.

"We've got six," Arnolfo said.

"Seven," said Aubrelia.

"Seven, then," he laughed. "From about five up to about twenty years old. I lose track. We've got to tend to the fields and animals, and that's what we concentrate on. We don't go too far from home like you and Juan."

"Well, it's a beautiful place to live," I said. "I love those hills."

***

Juan appeared at 8:00 a.m. I paid 500 guaraníes for my food and we walked to his house and then to a spring he had protected. We visited the *Centro de Salud* (health centre) and he showed me a pump he had installed. We visited a few springs he planned to protect and then walked to the far end of the community to a house where I was to eat and sleep.

A straw roof covered the two-room house with a breezeway between the rooms, which served as a dining area. A rough table and several chairs stood on the hard-packed dirt floor. Two women cooked over an open fire on the floor of a three-sided outbuilding behind the house. We ate pasta with beef and *mandioca*.

We talked until dark and the *señora* showed me to my bed in one of the rooms with her four sons. The two daughters slept in the other room with the parents. Before bed I made my way past the kitchen to the outhouse, a roofless, three-sided structure. The open side faced away from the house toward the brush. Split slabs of coconut tree trunks formed the floor over a deep pit.

I walked to Juan's in the morning. He was at church where he played guitar every Sunday. I hauled a few buckets of water from the spring to his house and took a splash bath. When he returned, we went to breakfast at a neighbour's and then we climbed a *cerro* (hill).

***

The peak of the rocky *cerro* stood a dozen stories above the main part of Potrero Arce where Juan lived near the school, church, health centre, and the *almacen*. We could see parts of the road winding into and out of the community, through wooded areas and fields slashed out of the woods. A breeze moved clouds across the soft blue sky. A half dozen parrots chattered in the shade. The ground sloped from the rocky top of the *cerro* to a dark green canopy of trees to slender, erect coconut trees to light green fields and glistening wetland. At a broad creek at the lowest point the land climbed up again to a smaller *cerro*. Near the creek and along the road smoke wafted from cooking fires beneath straw roofs. Cattle, horses and sheep grazed in clearings. Oxen pulled carts through fields. On all sides we saw distant, wooded, wet, unpopulated land.

We descended, dripping sweat, and drank water, bathed, and ate. Juan borrowed a cot and I spent the night at his house.

In the morning, I walked the eight kilometres to the highway. I changed my sweaty clothing and waited an hour for a bus that I rode to the terminal in Asunción, and then took another packed bus with some windows missing to Areguá. Passengers pressed against one another with no room to move. A hard rain fell and soaked those of us near the missing windows.

***

Back in Areguá, after the trip to Potrero Arce, three languages spun through my brain. We entered the sixth week of training. I had learnt so much about Paraguay, its languages, and my prospective volunteer work that I realized how little I knew. I had so much more to learn before I could even begin to make a difference in the lives of others. I'd been supplementing my training by dancing with the local folk dance group to learn Paraguayan dances and to practice speaking Spanish.

At training one day we split into small groups and went to nearby towns to conduct a community analysis. I went to Patiño, about five kilometres from Areguá, with two women. We went from house to house asking questions about water sources and water quality and community needs.

Residents received us well, invited us to sit, and offered us food and drink. We drank *tereré* at each stop. When we'd gathered sufficient data, we returned to the training centre.

The following morning, fourteen of us boarded a bus and rode four hours to San Juan, a town forty kilometres from Caaguazú. Seventeen of the kilometres were on a rutted, red dirt road. There we met another volunteer. We stayed two nights and built and installed two well pumps. We slept on the porch of the *Centro de Salud* and the floor of the police station. The training went well, but the accommodations were poor. To feast with us, residents slaughtered an old ox, the toughest meat I'd ever eaten. Only the avocadoes were palatable for most of us.

4

The time arrived for site assignments, where we would spend the next two years. I agreed to a site that had been offered to others in our group and groups before ours. No one had taken it because of a problem with *mal de chagas*, Chagas Disease, a tropical parasitic disease. The University of Science in Asunción was studying the prevalence of *mal de chagas* with a German agency, The German Technical Cooperation Agency (GTZ).

The *mal de chagas* disease results from a bite from an infected beetle called "*vinchuca*" in Paraguay and is known as a "kissing bug" or "triatomine" in English. The *vinchuca* lives in straw-roofed houses and hides behind wall hangings, such as pictures. The *vinchuca* bites a person on a soft area such as the face and immediately defecates in the bite. *Mal de chagas* can be fatal, usually because of heart muscle damage. Symptoms often do not occur for years after a bite.

I would organize a battle against parasites in my site, Guazú Cuá, and would promote improved housing to help fight *mal de chagas*, promote sanitary latrines, dental health, protected water sources, and would coordinate with the Paraguayan GTZ contractor, a doctor.

In 1988, a German agency, GTZ, began the study of *mal de chagas* in the area and as a result had financed the building of a ten room *Centro de Salud*. The building was the largest in the community.

Guazú Cuá, a cotton farming community of about 400

people, with no electricity or indoor plumbing, and fifteen kilometres off the main road with no public transportation, covers a twenty-five square kilometre area, eighty-five kilometres southeast of Asunción. I would be the first Peace Corps volunteer in the area.

On April 3, I was scheduled travel to Guazú Cuá for a week-long visit. My site coordinator arranged for me to stay with a family during the visit.

***

Meanwhile, in Areguá, I organized a dance for Wednesday night, March 29. I hoped for an equal number of Americans and Paraguayans. I hung a dance poster at the training centre and I asked Nanci, one of the Paraguayan dancers, to teach a couple of Paraguayan folk dances and I would teach a few dances as well. I visited Nanci at her parent's house to talk about the dance. She told me some things about Americans that I hadn't noticed.

She said that American women often walk alone with men. In Paraguay, a single woman walking alone with a man puts the woman in a bad light. Also, Americans often drink from bottles. Paraguayans drink from a glass or from a bottle through a straw. Several people may share a glass between them, but did not drink directly from the bottle.

Thursday morning, I walked to the training centre happy because the folk dance exchange had gone well. I'd taught a swing and a waltz to Cajun music. The Areguá natives had taught the Paraguayan Polka, a traditional polka to Paraguayan harp music, and demonstrated a dance done without partners, women only. Few Americans attended, because the night before sixteen people in my group had chartered a bus to go to a soccer game.

***

The next morning I stepped into the training centre and felt a palpable pall of pain descend. Two women sat on stairs weeping. A man sat alone, head down, his wrist wrapped in a bandage. As people trickled in Peace Corps administrators arrived and met with the CHP director.

Peace Corps suspended classes for the day. The group that chartered the bus had gone to the soccer game but the game lasted only fifteen minutes because of a problem with lighting caused by rain. On the way back to Areguá, the bus hit a median and flipped. Michelle Drabiski, a twenty-three-year-old nurse, died in the crash from cranial trauma: a fractured skull. Four other women went to the hospital with a broken foot, a broken collar bone, a concussion and one with possible head injuries. Five others were hurt but not seriously.

I heard a few versions of the crash. One said that the bus stopped at an *almacen* so people could buy beer and soda. It then continued toward Areguá for about five minutes before it went into an unexplained skid and flipped over the median. Michelle had been standing in or near the stairwell and fell out the door and under the bus.

Another account had the bus already equipped with plenty of beer and whiskey and a driver that may have been drinking as well as the passengers. They did stop at an *almacen*, and when the bus returned to the road it went too fast. The driver suddenly turned, skidded, hit the median, and flipped.

Steve, an *aspirante* a few years my senior, told me he was asleep on the bus and didn't know how it crashed. He remembered waking and walking out of the bus through the shattered windshield. Nearly everyone was cut by flying glass as the windows were not made of safety glass.

The agency suspended another day of classes. Michelle's parents arrived. Peace Corps provided grief counsellors. We attended a mass for Michelle at the church, *Nuestra Señora de la Candelaria*. Peace Corps Paraguay dedicated part of its library to the memory of Michelle Drabiski.

5

It was my first site visit. A woman sat on a low stool beside a tub in the shade of a mango tree near a well in front of her straw-roofed, wood-framed house. A cigar hung from her mouth as she hand-washed clothes. I clapped my hands at her gate to draw her attention.

"Can you tell me how to get to Guazú Cuá?" I asked, as I stood sweat-soaked and disoriented after a two-and-a-half-hour hike along a road of wet red dirt rutted from ox-drawn carts.

The road led from Sapucaí, Paraguay, over a ridge thick with coconut trees, lapacho trees, cotton fields, citrus trees, and occasional houses built of rough-cut planks with straw roofs and dirt floors.

I'd left Areguá at 5:00 a.m. and had taken three busses to Sapucaí, a town on the railroad tracks where the 1860s train repair shop operated. Paraguay had the first train to function in South America. In 1989, it continued to use a wood-fired steam train. Steam also powered the mainly British equipment in the repair shop.

From Sapucaí, I'd followed my hand-sketched map and began hiking along a dirt road that gradually climbed a *cerro*—Cerro Verde. I walked and my back dripped sweat where my small pack hung. An ox-drawn cart approached and I asked the driver if I was on the road to Guazú Cuá. The man grinned toothlessly and gestured here and there while speaking Guaraní. I understood not a word.

"*Gracias*," I said, and walked on.

At the top of Cerro Verde I'd met a man on horseback. I asked him how to get to Guazú Cuá. He gestured to follow him and he started along a trail. He stopped at a rock outcropping and pointed to some buildings several kilometres past the bottom of the *cerro*. Guazú Cuá, he said, and spoke to me in Guaraní. I understood nothing but I knew I had some ground to cover to reach my destination. I walked back to the road with the man, thanked him, and continued on my way.

The south winds of autumn bring frequent rain and the sometimes obscure road remained wet from the last rain. After I'd been on the road from Sapucaí for almost three hours I came upon the woman washing clothes.

"Where is Guazú Cuá?" I said.

She looked at me as if I'd appeared from a strange, faraway land—which, essentially, I had. She took the cigar from her mouth and smiled with no front teeth.

"*Acá nomás*," she said.

I translated that to mean "*here, no more*." It's not here anymore, I thought.

She said something in Guaraní and gestured with her cigar down the road. I thanked her and continued walking. I wondered if travelling to a community that no longer existed, according to my interpretation of *nomás*, was part of the training. Could my site be a ghost town, a farming community, that didn't make it? The assignment entailed, though, staying at the site for five days. It had to exist. It was, after all, where I would serve as a Peace Corps volunteer for two years.

Up ahead, I saw a horse trotting toward me with a woman astride wearing a wide-brimmed hat, barefoot, her waist-length black hair undulating with the rhythm of the horse. Was I delirious in the heat? When she reached me she reined in the horse, looked at me with smiling coffee brown eyes, and said, "Marcos?"

Lucilla, an aide at the *Centro de Salud*, had been expecting me for this site visit and led me to the home of my host family. The home consisted of two rooms with a lean-to on one side that served as a kitchen, and a roofless outhouse twenty metres away.

The Gonzalez family, six men and four women, lived in the house, although most of the men worked away and were home only a few days a month. I slept in one of the rooms and the rest of the family slept in the other. I paid 500 guaranies a day for room and board.

The Gonzalez home was located in the centre of Guazú Cuá on a corner. Across the grassy, dung-strewn street sat the school. Across the street on the left was the *capilla*, or chapel. On the right side of the house, Antonio and Vilma had an *almacen*. Vilma was also the director of the school. Beside the *almacen* was the new *Centro de Salud* that GTZ had built, and across from the *Centro de Salud* was the police station, where three soldiers were stationed. Cattle, oxen, and horses grazed in the streets. Chickens roamed free pecking for food and pigs rooted in the grassy fields. A woman passed with a five-gallon bucket of water on her head.

Two of the Gonzalez girls were seventeen-year-old twins, Ada and Eda. They showed me around the community centre. I met Antonio at the *almacen* and I told him I wanted to learn to ride a horse so I could buy one later. He stood barefoot in the mud and listened but didn't understand my Spanish and the girls translated in Guaraní. He yelled to someone in Guaraní and a few minutes later his son handed me the reins of a saddled horse. Antonio grabbed a tattered, wide-brimmed straw hat and put in on his head. He untied his horse from where it was tethered near the fence and he mounted and gestured for me to follow.

We rode the perimeter of the central area of Guazú Cuá, which consisted of lots measuring about thirty by fifty metres with an area of *campo común*, common ground where all residents could graze sheep, horses, and cattle. Outside the central area the lots contained twenty-five hectares, or about sixty acres. Residents occupied and cultivated these lots, usually with cotton and beans.

After the ride, I returned to the Gonzalez home. I grabbed a bucket to get some water from the well to bathe, but Ada and Eda wouldn't allow it and fetched the water for me. I couldn't do anything for myself. As soon as I took off my shirt, one of the girls grabbed it and started washing it.

I took a bucket bath in the *bañadera*, an open roof rickety structure of coconut bark lashed together loosely with vines and a wobbly wood floor elevated on rocks with a torn, burlap sack for a door. After, *Ña* Gonzalez served dinner, the worst I'd had since I'd been in Paraguay. It consisted of watery rice, dried meat like jerky, and *mandioca*—and that's all we ate during my stay. Not a vegetable did I see. I wanted to use a napkin, too, but when one finished a meal, the edge of the tablecloth served to wipe one's mouth.

Rain fell that evening and I had to relieve myself of my dinner. I took my flashlight and manoeuvred through the slick mud to the three-sided outhouse. The coconut bark walls stood chest high. A cloth secured privacy on the open side. Split coconut trunks with a hole cut in the middle formed the floor which bent to my weight. A pile of wet corn cobs in a corner substituted for toilet paper. I shined my light and saw that water trembling with thousands of squirming white worms filled the pit to the floor. I dropped my shorts, squatted and looked into the falling rain and realized how alienated from nature I had been.

***

I spent most of my five days riding horses and meeting people. At the school I spoke with a teacher.

"It's nice to meet you," Andrea said, shaking my hand. "Marcos, right? We heard you were coming. People here don't want to change. We need your help."

And at the health centre I met a few women sitting on a bench in the shade, waiting for a volunteer to open the office.

"What's it like where you're from?" one asked, looking past me. She gestured toward a grapefruit tree and said, "Do you have those there?"

"Chickens? Those?" another said, pointing to a couple of skinny chickens scratching in the dirt.

Three children followed me as I walked, staring and laughing and speaking Guaraní. A man sat at the edge of a wood, his head bobbing, drinking from a Coca-Cola bottle half full of a clear liquid. He exchanged words with the children and then held his bottle up. "Do you have this in your country?" he said.

I also met a neighbour, a slender, coppery young lady named Noemi with long black hair and a baby boy in her arms, Augusto. Her maternal aunt and other relatives from the paternal side lived in Potrero Arce and they all knew Juan, the volunteer I had visited. I walked next door to Noemi's house to meet her family. They also had a store, but sold mostly beer and soda. An old tattered billiards table with warped cue sticks sat on a brick floor beneath a straw roof in front of the window where the beer and soda was dispensed. I took a few pictures and went back to the Gonzalez house.

One day I sat in a chair in the yard at the Gonzalez house. A cart passed drawn by four oxen. In the cart women in black wailed eerily over a small casket, which held the body of a five-year-old boy who had been bitten by a snake the day before when he reached into some brush to retrieve a soccer ball. The rest of the family followed the casket on foot and on horseback. Ada and Eda gestured to me to follow them and we joined the procession. We stopped at the *capilla* and men unloaded the casket.

The *capilla* was open with a tile roof. Only at one end hung doors that opened to an altar with a statue of the *Virgin de Fatima*, the patron saint of Guazú Cuá. Men placed the coffin before the altar and one led the group in prayers. They closed the doors to the altar and loaded the coffin into the cart. We walked toward the cemetery.

Women wailed and men wept as we trudged through mud, pasture, and across a creek to the gate of the cemetery. About 100 people, most of them barefoot, gathered around a freshly dug grave. The incessant wailing filled the air with the heaviness of reality, of the now, and the sound chilled me through. I blinked and wiped tears from my eyes. We said more prayers and then men used ropes to lower the rough-hewn casket into the wet earth. The boy's mother, one of the women in the cart, scooped dirt from a pile and threw it into the grave. One by one people threw dirt into the grave. When all had done so, we left and the boy's father stayed behind alone to finish burying his son. On the way out I noted that many of the graves were those of children.

***

Every other Friday a doctor and a biologist visited the *Centro de Salud*, providing the road was passable. This week they arrived in Guazú Cuá in a white, four-wheel drive Mercedes Benz. The doctor, Dr N_____, in polished shoes and starched office garb, gold on his neck and wrists, nonchalant with slick black hair and dark, conniving eyes, had a contract with GTZ. I rode out of the site with them and got permission from Dr N_____ to live in the *Centro de Salud* when I returned to the site after training. I spoke mostly with the biologist. The doctor appeared distant and I suspected that he considered me his employee. The biologist told me that thirty percent of the population of Guazú Cuá tested positive for *mal de chagas*. That was exceptionally high, since the norm was about 0.03%. That justified the GTZ contract and the *Centro de Salud*.

We bounced, slid and spun through muddy ox cart ruts for fifteen kilometres until we reached the main road, which was also dirt. Fifteen kilometres later we hit the asphalt of Route 1 in Paraguarí. Dr N_____ stopped at a restaurant and bought *empanadas*, *mandioca* and soda for the three of us. We ate as we rode toward Asunción. When we finished eating Dr N_____ asked us to put our trash in a bag. He took the bag, rolled down his window and tossed the trash onto Route 1.

***

The last week of Peace Corps Paraguay Volunteer training, I went to Asunción to meet with Dr N_____. I found his brick office building on a hill down a cobblestone street. A tree with a broad, leafy canopy grew in the middle of the street and its roots pushed up rough cut cobblestones. Grapefruit trees shaded a broken sidewalk. A dark liquid trickled along the gutter between stones. I climbed a tiled staircase the office on the second floor.

Dr N_____ sat behind his desk groomed, debonair, a man of business and as unlike a doctor as the last time I'd seen him.

"I want to see you promote environmental sanitation, Marcos," he said.

That was exactly what I wanted to do. I feared he would expect me to work under him.

"I'm here to help you complete your mission for Peace Corps, helping with the *mal de chagas* project," he said, as if he held my success in his hand. He gave me his contact information and I left feeling good about my site, yet wary of my counterpart, the doctor. I'd been training to work in environmental sanitation and not with *mal de Chagas*.

I wouldn't go to Guazú Cuá to live until after the elections on May 1, because Peace Corps wanted to take precautions just in case something happened. Nothing did happen. General Rodriguez, who had led the coup in February, won the election with seventy-five percent of the votes.

***

In a meeting room at the training centre in Areguá, we had a speaker explain the situation in Paraguay in his opinion. The young, thin, red-haired man lit a cigarette and apologized for his poor English and subsequently spoke in very good English, smoking almost non-stop.

He told us that Stroessner himself gained power through a military coup in 1954. The international community accused Stroessner's government of human rights violations throughout his reign. Alleged abuses included the detention of dissidents without trial, the disappearance of political opponents, and the use of torture. The Colorado Party backed Stroessner and he reciprocated by allowing party leaders to monopolize and to profit from corruption and contraband. The Paraguayan speaker told us a bit of the recent history of the Republic of Paraguay. President Stroessner had lost control long before General Rodriguez and his fellow conspirators overthrew him in February 1989, he explained.

The speaker had just finished his summary when the time was up.

***

On Friday, April 28, 1989, the end of the last week of Peace Corps Paraguay training, the United States Ambassador to Paraguay swore us in as Peace Corps volunteers. First he spoke. Next the Paraguayan Minister of Health spoke, and then the Director of CHP. Finally, Don Peterson, Director of Peace Corps Paraguay, spoke. He teared up while talking about Michelle (the nurse who died in a bus crash during training). Each speaker mentioned Michelle. One of the women in our group gave a speech in Spanish and one of the men in Guaraní. Finally, a local politician spoke.

Each of us went to our Areguá host families and presented them with a rose. Two women presented a board covered with pictures of our group to the staff. Then two women presented gifts to each of the teachers.

Twelve weeks of training in Areguá had ended. Thirty-three of the original thirty-five members of the group remained. We were twenty-two women and eleven men. Twenty-one had recently graduated from college. Eight joined Peace Corps after retirement. Four of us left careers to serve.

We had going away parties every night. On Wednesday night, May 3, *Ña* Fidencia prepared a special dinner for me: chicken, pizza, *sopa paraguaya* (a corn bread with cheese and onions), *mandioca*, and drinks. The entire family attended. I went to bed before midnight.

At 2:00 a.m. I awoke to the voices of six men, family members, outside my window singing in Spanish. They stumbled through three songs. I got up and went and sat outside with them. It was my goodbye serenade from them at the end of the last week of Peace Corps Paraguay volunteer training.

The next night I packed for my move to Guazú Cuá. I left the house at about ten the next morning after my goodbyes with *Ña* Fidencia and *Don* Pedro with promises to return often to visit. Ambivalence about leaving overwhelmed me. The house had become my home, but I hadn't gone to Paraguay to live in *Ña* Fidencia's house for two years.

I walked to the bus stop and caught a bus to the terminal in Asunción. I bought a short wave radio and then I boarded a bus to Paraguarí. I found the *Proyecto Paraguarí*, a development project, and met my contact, Luci. She showed me to a room there, a cell-like structure with a cot and without other facilities except a bathroom down the hall. She then showed me where she lived, about a block away in an apartment on the third floor. It had been raining and we wouldn't be able to get to Guazú Cuá for a few days. Luci or one of her co-workers would drive me to Guazú Cuá with my luggage and furniture when the road dried.

I settled into my room at the *Proyecto Paraguarí* and then walked around the City of Paraguarí and found a place to eat. Route 1 bisects Paraguarí, but most of the streets within the city of 30,000 are cobblestone or dirt. Route 1 veers south at Paraguarí and runs to Encarnación on the border of Argentina. Paraguarí supplies the surrounding farming communities and buyers from Paraguarí also look for cotton arriving from the country.

The next day I went to a carpenter recommended by Luci. I ordered a *ropero* (40,000 guaraníes), a table (15,000 guaraníes), a bed (25,000 guaraníes), and two chairs (5,000 guaraníes each) for the room I'd have in the Guazú Cuá *Centro de Salud*. The exchange rate at the time was about 300 guaraníes per dollar. It would take a couple of days to construct my order.

I walked the market and bought a frying pan and a pot, some cups, and a wool poncho, trying to barter for each item. The vendors, though, didn't seem to care if I made a purchase or not, so I paid the asking price. I already had a mosquito net I'd brought from the United States. In the long run, the wool poncho served as the best and most useful buy.

Luci invited me to a party. I went with her and met Rosa, the woman who would drive me to Guazú Cuá. Rosa worked for *Proyecto Paraguarí*. Both Luci and Rosa held degrees and were single, but they were in their thirties. That, so I'd heard, was spinster age for a Paraguayan woman at the time. At the party we danced and talked. Rosa drove me home because she had the company truck, a Toyota.

A few days later, Rosa picked me up to get my furniture and to drive me to my site, where there was to be a bull fight. We drove to the carpenter for my furniture, but we couldn't fit the table and chairs in the bed of the truck, so Rosa said she'd get them to me in a few days. We left the carpenter and bought fuel for the truck and then bought some *empanadas* and *mandioca* to eat on the way. We left Paraguarí on a red dirt road toward Escobar, a town fifteen kilometres north of Guazú Cuá. At Escobar Rosa stopped at the *Centro de Salud* there and got another man to ride the rest of the way with us so she wouldn't have to ride back alone. He took the wheel.

We turned south just after Escobar onto a rocky, rutted, muddy road. As we bounced and slid I looked at the coconut trees, cotton fields, sugar cane fields, at the patches of *mandioca*, and at the men and women working the fields with hoes and oxen. We passed wood houses with straw roofs, and we saw burros, horses, cows, pigs, sheep, chickens, and people. We buried the Toyota in black mud and a farmer pulled us out with a team of oxen.

While we were stuck, an ox cart passed full of white bags filled with small hard rolls. A bag bounced out of the cart and splashed into a muddy puddle. A woman following the cart snatched the bag out of the puddle and grabbed a fresh bag from the cart. She dumped the good rolls into the fresh bag and left the wet ones in the dirty bag, probably to feed to pigs.

A couple hours after we left Escobar we parked in front of the *Centro de Salud* in Guazú Cuá, the biggest building in the community except for the school. A wide, tiled, covered breezeway separated the two sections of the white, plastered brick with a tile roof. The breezeway served as a waiting area and a community meeting place. One side of the building consisted of an office and two rooms where a doctor could see patients. The other end of the building had three rooms and bathrooms with showers, but no running water. All the rooms had glass windows, the only ones in the community. These rooms would be reserved for those who needed to be hospitalized. One end of the breezeway doubled as a loading dock. In front of the hospital end of the *Centro de Salud* were public outhouses. A desk, a few chairs, and a locking metal cabinet furnished the office. The cabinet held medical supplies such as aspirins, antibiotics, a blood pressure gage, blood pressure medication and tapes and dressings. Gaspar Meza ran the *Centro de Salud* and was to be my counterpart along with Dr N_____ , who made occasional appearances. Often he sent another in his place.

Gaspar and a friend sat on a bench in the breezeway drinking *tereré*. Gaspar sported polished shoes and clean, ironed pants and shirt. A gold cross hung from his neck. His friend had stopped by from the field, barefoot and earthy. They gawked at Rosa.

Rosa introduced us and her friend and Gaspar's friend helped me unload the truck and place my furnishings in the last room of the hospital end of the building. I picked that room because there was a door at the end of the hall that exited to an area with a double sink with no water. There was a shallow, common well of non-potable water fifty metres from the sink.

Once we'd placed my furnishings, Rosa said she'd be back in a few days, weather permitting, with the rest of my furniture. She left with her friend from Escobar driving.

***

Gaspar came into my room to see what I had. "How much was the bed?" he said. "How much was the *ropero*?" he continued until he knew something about everything I had.

"Is this your full-time job, Gaspar?" I asked.

"I'm a nurse aide and the manager here. GTZ trained me," he said.

"So you're open here all day?"

"No. I come for a few hours and if no one needs me I go home. People know where to find me."

"You work for GTZ or Dr N_____?"

"I work for Dr N_____, but my pay comes from GTZ," he said.

"What about Lucilla?"

"She's a volunteer. GTZ trained her, too. She stops by a few times a week. What do you expect to do here?" Gaspar said.

"Promote environmental sanitation. I thought you were my counterpart."

"We *are* counterparts. I'll show you around once you get settled in. Do you have a place to eat?"

"Not yet."

"Let's go to my house. You can eat there for today."

He gave me a key to my side of the building and I locked the door. He locked the door to the office. His friend had disappeared. He walked behind the *Centro* and caught his horse and saddled it. He donned a wide-brimmed, blue cloth hat and gestured for me to follow. I walked beside his horse to his parents' house, a ten-minute hike.

Gaspar's mother served us a soupy dish of noodles with chunks of tough, stringy meat and *mandioca*. While we ate, she squeezed lemons and made a pitcher of lemonade.

"I'm going to take a *siesta*. I'll show you the community tomorrow morning. Meet me at the *Centro de Salud*."

"Okay. I'll be there anyway, since that's my home for now," I said.

I left Gaspar's house and walked toward the *Centro de Salud* alone. That was the only time I went to his house by invitation. I did go once more for my community census.

I walked to the school to look at the well and pump. A line ran from an elevated tank above the well to the health centre, which would have been the only source of running water in the area. The hand pump was missing a cap for the clean out hole and a gasket rotted causing the pump to leak. I found a chunk of wood and whittled it to size and plugged the clean out hole. I pumped for about an hour and raised enough water to the tank to bathe. Most of the water sprayed out from the rotted gasket and when I'd finished pumping I stood wet and in mud. Antonio sat in front of his *almacen* drinking *tereré* and watching me. I waved to him on my way to the *Centro de Salud* where I planned to enjoy my indoor plumbing before the water dripped out of the elevated tank.

6

The next day I went to look at the pump again, when I saw a pickup truck coming out of the woods and into the open area of Guazú Cuá. It wove and bounced until it reached the school and then stopped. Two men got out, one from *Proyecto Paraguarí* and the other from Ciba Geigy. He was in the area to collect money from the Cotton Growers Association for chemicals he sold on credit. He spoke some English.

"People want to know what you're going to do here," he said.

"Good question."

"It would be good if you'd work with the members of the Association. Why don't you come with us to the meeting?"

"Where?"

"Right there at Antonio's."

"Okay. I'll be right there," I said.

They got into their truck and drove to Antonio's and I followed on foot.

The street in front of Antonio's *almacen* stayed black and wet with the rich dung from cows, oxen, horses, pigs, and chickens and the urine from the same. Animals and people gathered there, and men tied their horses to the fence and others left their oxen and carts in front. Planks and rocks formed paths from the road through the dirt or mud of the front yard, depending upon the weather. Antonio's house consisted of three main buildings, the *almacen* and sleeping rooms, the kitchen, and a storage building. The storage building had been the original house with a straw roof and walls of vertical planks painted powder blue. The roof extended over a concrete walk, which is where people sat on rickety chairs drinking *tereré*, beer, or *caña*. The main plastered brick building faces the storage building and a porch roof connects to the blue building.

The window through which Antonio sold goods opened to the porch. The *almacen* itself was about two by three metres of shelves of wine, flour, sugar, rice, pasta, canned food, toilet paper and other such necessities, as well as a kerosene powered refrigerator. There was a shallow well and a bucket with a rope on the ground in the front yard. Behind the buildings stood the partially open kitchen, and beyond that, the outhouse and *bañadera*.

Antonio went barefoot with his pants rolled half way to his knees. He wore a shabby, dirty blue, wide-brimmed hat. He passed to and fro serving his customers.

Several men gathered and sat drinking *tereré* and and smoking hand-rolled cigars. They'd been in the field and wore *zapatillas* or went barefoot. All wore broad-brimmed hats and had soil from the fields on their calloused hands and feet. I shook hands with everyone and took a seat.

The Ciba Geigy representative presented his bill and an argument ensued in Guaraní. I didn't understand much but I gathered the men found an error in the amount of the bill. The representative agreed to a lesser amount and the meeting broke up just as Ada came looking for me to have dinner at her house.

First I returned to my room for some pictures I'd taken of the family during my first visit to Guazú Cuá. I also had a kilo of *dulce de guayaba* that Ña Fidencia made as a gift.

As I scraped the concrete-hard rolls with my teeth and

masticated leather-tough meat, the family looked at the pictures and ate the *dulce de guayaba* from the jar with a spoon. Noemi stopped by and I gave her pictures as well.

"How much are they?" she said.

"Nothing."

"*Muchas gracias.*"

I went into the cooking area to speak with *Ña* Gonzalez about a meal plan.

"*Ña* Gonzalez, I need a place to eat while I'm here. How much would you charge me per month to eat here?"

She turned her eyes up at the blackened straw roof and at dried beef strips hanging from a string stretched between two posts, and she stared into the fire on the dirt floor and at a blackened pot and then she locked eyes with me.

"I'll give you meals and do your laundry for 40,000 guaraníes a month."

I learnt in training that it should have been half that.

"That's more than I can afford, *Ña* Gonzalez."

"35,000 guaraníes, then."

"Okay. I'll pay you for the rest of this month." I calculated the cost from May 11, the following day. I counted out 23,000 guaraníes and handed them to her.

On my way back to the *Centro de Salud* I realized I'd have to find something better for June. Her food sat like gravel in my stomach. I shouldn't have made the deal, I thought. I knew she had taken advantage of me and I let her. My living stipend was 133,000 guaraníes a month, which included my bus fare to Asuncíon once a month and the expense of a two-day stay there.

The next morning, I went to look at the pump again. Someone had pulled my plug. I went to the shallow well near my room and hauled water to the bathroom to wash and to flush the toilet.

Outside ox carts packed with white bags of cotton gathered at the loading dock. I walked out to look. The Association would use the dock to store cotton and would have a truck haul it all to market at the same time. Twenty-two families belonged to the Association, about a quarter of the community. Most hauled their cotton to market in ox carts.

Men weighed sacks, numbered them, named them and stacked them on the loading dock. I tried to help, but the growers spoke Guaraní and cast glances of mistrust toward me. Cotton provided the only cash they would see in a year. They argued about leaving the cash crop in a public place.

Out in the common pasture, the *campo común*, I saw a group of men constructing a fence. I left the Association and walked ten minutes across the pasture to the workers while I looked at the thin Bhrama cattle and dirty sheep grazing. The men had ox carts loaded with bamboo and slender wood posts and a few planks. They built a circular corral about twenty metres in diameter and two metres high with the posts and split bamboo woven between the posts and tied with vines. They used stronger material to build a chute and a holding pen.

"Are you going to brand cattle?" I asked.

"It's for the *torine*, for the festival of the *Virgin de Fatima*," a worker said.

"*Torine?*"

"*El toro. Vamos a hacer correr a los toros.*"

I gestured like a matador and they laughed and shook their heads.

Saturday there would be a fiesta. I crossed the pasture to the police station, the *Alcaldía*.

"How do you feel here so far?" the soldier said.

"Not bad, but lost."

That night I sat in the breezeway of the *Centro de Salud*. I saw a few candles flickering in distant houses and the constant blinking lights of fireflies. A quarter moon drifted in a clear sky brilliant with stars.

In the morning a convoy of ox carts delivered cases of beer to the *Centro de Salud* for the fiesta. I helped Antonio stack the cases in an empty room in the *Centro*. The fiesta would be held at my place.

I waited for Vita, my site coordinator, to arrive to formally introduce me to the community. But she didn't come, and neither did Rosa with my furniture. I sat with some men from the local cotton association and drank *tereré*. Cotton farmers had formed an association to bargain as a group with buyers.

"I want to go to every house to meet the people and find out how I can help them," I said.

"There will be plenty of pretty young girls at the fiesta tomorrow," one said. And they taught me what to say in Guaraní to a girl, should I want one.

Saturday night people scrambled about the breezeway at the *Centro de Salud* setting up tables to sell drinks and food, and other small tables for card games. A cassette tape player connected to speaker boxes provided the music. Benches lined the perimeter of the front yard of the *Centro*. Girls arrived in dresses with their high-heeled shoes in their hands and they put them on when they sat. A pair of women accompanied each girl, usually a mother and aunt or grandmother. Men tied their horses to the fence and stood near the beer table in their polished shoes and broad-brimmed hats. They drank and joked and gestured toward the girls. Others played cards.

Two young men approached girls. They spoke first to the chaperones and then led the girls to the dance area on the grass. Soon many danced and at midnight the crowd drifted off except a couple of drunks who sat outside yelling and drinking. I listened to them from my bed.

The next morning began with a mass in the *capilla*, given by a priest from *Colonia Santa Isabel*, a leper colony near Cerro Verde. He came once a year for the festival of the *Virgin de Fatima* and to baptize babies, the only opportunity during the year to be baptized. The priest lectured about motherhood and the importance of marriage to a dozen women—half single mothers—each with a baby. Noemi was there and so was Sofía, one of the Gonzalez girls. They formed a semi-circle around the priest and he walked to each baby, mumbled something and poured a bit of water over each head.

Sofía motioned for me to take a picture as no one else in the community had a camera. I snapped a photograph and then all the mothers wanted a photograph and offered to pay. I snapped a dozen or so and then cut them off.

As I took photographs, men paraded around the *capilla* with the statue of the *Virgin de Fatima*. Fireworks exploded. I went home to wait for the bullfight.

7

When I walked to the corral, I saw a grandstand beside the corral built with posts, branches and bamboo latched together with vines and strips of leather with rough planks for a floor. Twenty people stood on it, but I wouldn't have stood on it with more than four. People surrounded the corral pushing against one another to get a closer view. Vendors sold *chipa* from baskets. *Chipa* is a traditional bread baked with *mandioca* flour, wheat flour, eggs, and *queso paraguayo* (Paraguayan cheese).

A table behind the grandstand held cases of beer and soda and bottles of *caña*.

A truck arrived loaded with seven bulls and backed up to the chute. *Vaqueros* roped one bull and pulled it into the chute and sawed off the tips of his horns.

Three men walked around inside the corral, two dressed as clowns and one as a *torero*. The gate opened and the Brahma bull charged the men and they teased it with capes. The *torero* grabbed the bull by the tail and twisted it. One of the clowns ran directly at the bull and planted his hands on the bull's head and knocked the bull off its hooves. They worked all seven bulls that way.

In the chute, *vaqueros* tied a short rope to a bull's right hind leg and a longer rope to its right front leg. The bull already had a rope around its neck. When the gate opened the bull charged but stopped short because of the rope on its hind leg. A *vaquero* jerked the rope on its front leg and the bull fell. Six *vaqueros* jumped on the bull to hold it down while the bull snorted in the dust, its chest heaving with quick breaths, its taut muscles straining against the captors. One tied a sash around the bull's chest and back.

The *torero* got a hold on the sash and straddled the bull. A *vaquero* untied the ropes. The six sprang from the bull and the bull leapt to its feet bucking and running.

The *torero* rode until he was thrown and the bull turned and stomped on him before the clowns could divert it.

Some bulls charged into the ring and stopped and stared and urinated or defecated. One seemingly passive bull suddenly caught a clown with its bloody, sawed off horns and threw the clown across the corral.

Vaqueros tied a wad of guaraní notes to the scrotum of one bull. Anyone could enter the ring to yank the wad of bills from the bull's scrotum and keep the bills as a prize. The chute gate opened and the bull charged into the ring, the bills swinging from its scrotum. He stopped, looked around and ran straight for the corral fence. Spectators screamed and scattered and laughed and the bull splintered the bamboo fence on its way through and ran bucking into the woods, the guaraní notes swinging. *Vaqueros* on horseback chased it.

Rain fell and the crowd dispersed. I went to the *Centro de Salud* to wait out the rain, but it rained from a dark sky for two more days, leaving the grassy road to mud and puddles and the creeks surging at their high banks. Runoff cut new rivulets on the way to the creeks.

I tuned in my short wave radio and heard that General Rodriguez took the oath as the President of Paraguay 101 days after the coup. The Peronist Party won the presidential election in Argentina. The United States was trying to throw Noriega out of Panama.

Antonio invited me to eat at his house and we had *borí borí*, a soup-like dish with corn balls and beef. We ate it with *mandioca* and the traditional meal was one of the best I'd had.

"I'm going to build a garden at the *Centro de Salud*," I said.

"Make sure you get enough fertilizer from the woods," Antonio said.

"I will, and then I'm going to conduct a community census. I'll go to all the houses."

"People want to know what you're doing."

"I know. I'll explain to them that I'm identifying community needs so I know where to begin, and I'm going to begin with my own garden."

"Good luck. Let me know if I can help you."

"Thanks, Antonio."

8

I tired of digging and turning soil behind the *Centro de Salud*. I sat in the breezeway in shorts and a tee-shirt gazing at the full moon and the blinking fireflies. I decided to walk out of my site to take a break for a couple of days and to visit my counterpart, the doctor. I felt as though the community watched me like a carnival attraction.

People stared and spoke about me when I passed, and when with them they'd talk about me in Guaraní, or so I assumed. I hadn't learnt enough Guaraní to be sure.

"What's he going to do?"
"Why does he live alone?"
"Does he like women?"
"Why isn't he married?"
"Why would he want to come here from a rich county?"

I packed a change of clothes and locked my side of the *Centro de Salud* and walked across the field to the road out of Guazú Cuá with the doctor on my mind. I walked for two and a half hours past cotton fields, through woods, past houses, over hills and across creeks. I walked beside the deep ruts cut by ox cart wheels.

At each creek, the current flowed through a concrete culvert. The flat top of the culverts formed a road bed, but nothing connected them to the banks of the creeks. It was a road improvement project started but never finished.

Vehicles, ox carts, and horses that used the road forded the creeks.

I passed through Chircal, a community about the size of Guazú Cuá, and half way to the main road. Chircal spread along a wooded, rocky hillside. Farmers slashed and burnt the woods to create space to plant cotton and other crops. I crossed a ridge at Chircal and walked downhill the last several kilometres to the road. I stopped to wash at the last creek before the road. During the two and a half hour walk, I'd passed two men on horseback and seen several farmers working in fields. Houses looked empty except for the smoke of cooking fires. I heard my steps on the soil, the water rippling through culverts and churning around rocks. I heard the trills of mocking birds, screeches of parrots, and I watched the antics of killdeers protecting their nests. Scrawny cattle grazed in fields and ripped leaves from tree branches.

A house at the intersection of the road doubled as a bus stop where travellers could wait in the shade or out of the rain and purchase food and drinks from the home owner. The bus passed every two hours when the road was dry. The road was closed when it rained.

I waited with a dozen people, all with cloth shopping bags, and all speaking Guaraní. We heard the blast of a bus air horn so we scrambled out to the road. A blue bus skidded to a stop, throwing its passengers about. Two people made their way off the bus and others passed their bags out the windows to them. Men rode on the back bumper, and some stood on wheel wells and held on to something sturdy inside the bus. Riders crammed both stairwells and others hung outside the stairwells with barely a toehold. The driver jammed the bus into gear and lumbered off, leaving most of us at the bus stop. I started walking the next fifteen kilometres to Paraguarí and the asphalt of Route 1.

I walked toward Escobar with the rocky, wooded *cordil-*

lera (a chain of mountains) rising on my right and a coconut tree-dotted pasture spread to my left. The red dirt of the road clung to my shoes and pant legs. A pickup truck stopped for me and I jumped into the bed and rode to Paraguarí.

I found Luci at the *Proyecto Paraguaraí* and had lunch with her. I told her about my site and about how I felt it difficult to acclimate.

"You just got here," she laughed. "Give it several months. You won't want to leave, especially when you meet a *kuÑataí porá* (a beautiful, single girl)."

She walked with me to the terminal where I boarded a bus to Asunción and where I took a room at the *Itapúa Residencial* for 4,000 guaraníes a night including breakfast.

I made a list of things to do. I wanted a map of Guazú Cuá and Luci had given me the address of the *Ministerio de Hacienda* (the Treasury Department) and a contact name. I found the building near the Plaza Uruguaya and the train station in Asunción. Two open-air bookstores operated on corners of the plaza. Hotels, stores, and office buildings lined the streets around the area. Trees shaded the entire grassy square. Shoe shine boys in flip-flops flipped coins against the base of a statue. Women in short skirts and tight tank tops winked and gestured to me as I crossed the square.

I entered the building and asked for the woman. She came and took me to her desk in a room with high ceilings with fans hanging from every rafter. A dozen desks and women and typewriters occupied most of the space on the wood floor of the second storey of the building. All the windows were open. Most of the women stopped typing or filing to look at me.

"Luci from *Proyecto Paraguarí* gave me your name. She

said you could get me a map of a *compañía* of Escobar, Guazú Cuá."

"No problem. I can get you a copy of the original plat for 30,000 guaraníes."

"I can't pay that much for a map."

"You can't get it anywhere else, though. It's a valuable map."

"I'm sure it is, but that's too much."

"I might be able to get it for 20,000, but it might not be as good a copy."

"That's still a lot more than Luci told me I'd pay."

"Let me talk to a friend. Maybe we can work something out." She left me at her desk and I smiled at the women looking at me.

"You're in luck. We can get it for 10,000."

"Okay. I'll take it."

"Give me the 10,000 and wait here. I'll be right back."

A half hour later she returned with a blueprint of Guazú Cuá, which showed the boundaries, water courses, lots and dimensions, and the roads, even though they hadn't been built. I'd use it for my community census and as a guide to draw my own map on Mylar. Luci told me the plat would cost about 3,000 guaraníes.

"Anytime you need help, just stop by."

"I will, thanks."

I dropped off my film to be developed and then stopped by the SENASA office, the national Environmental Protection Agency. I introduced myself as a Peace Corps volunteer working with a doctor near Escobar and a secretary led me into a courtyard and up a staircase to an office where a few men sat about drinking tereré and reading newspapers. One man passed the *guampa* to me and I sucked the cold drink through the silver *bombilla* and passed it back. I pulled a slip of paper from my pocket with a description of the school pump and asked about getting spare parts.

"It's a Brazilian pump and if you want parts you'll have to go to Brazil. It would be easier to get a new pump," one of the men explained.

"Let's see that map," another said, pointing at my rolled map.

I put it on a desk and unrolled it and explained where I bought it.

"How much did you pay?"

"10,000 guaraníes," I said, "but the woman wanted 30,000 at first."

"She took advantage of you." They laughed. "It's just a copy. Should have cost a thousand or two."

"If there's anything we can do to help you, just let us know."

"I will, thanks." I shook hands all around and headed for the hospital where GTZ had offices. I wanted to get a ride back to my site when the doctor made his next trip. Vita had told me that he would visit the site every Friday, weather and road conditions permitting. With a ride I wouldn't have to wait for busses or walk for hours to my site.

I found no one, so I went to Dr N_____'s office. I introduced myself and waited an hour and a half to speak with him.

While I waited I wondered about my relationship with this doctor. I had expected my counterpart to be someone I would work with, not wait for. So far, Dr _____ had acted as if I were a subordinate, but I wondered if this might just be his personality. I'd only met him a few times and my impression may have been wrong. It did help that he lived and worked in Asunción. I wanted to ride to Guazú Cuá with him and his team.

When Dr N_____ appeared, I asked about riding with him.

"Meet me at the traffic circle on *Calle Ultima* near *Ñu Guazú* Park at 6:30 Friday morning," he said. "I'll have a couple of doctors and a nurse with me, but we'll squeeze you in."

***

On Friday morning, I paid my bill at the *Itapúa Residencial* and tried to catch a bus. Each one passed crammed with commuters, so I hailed a cab. The driver started his meter and sped toward the traffic circle and pulled into a gas station.

"*Cinco mil*," he said.

"5,000? Your meter says 3,500."

"It doesn't work."

"Why did you set it?"

"*Cinco mil, por favor.*"

I handed him a 5,000 guaraní note and got out of the cab to wait for Dr N_____ and his team. I bought a couple of *empanadas* and a coffee from a woman with a rickety table under a sun-bleached umbrella. I watched the stream of busses, cars, motor scooters, street vendors, and commuters while I waited in the shade of a tree. At 8:00 I spotted the white, four-wheel drive Mercedes Benz, so I ran to the curb and waved.

Dr N_____ explained his plans for Guazú Cuá as we drove.

"GTZ split Guazú Cuá into eight sections and each section has a local volunteer leader. I want a model garden at the *Centro de Salud* and one in each section."

"I'm working on a garden with good soil at the *Centro de Salud*."

"Good. I want one at the school, too. You'll work in each section talking to the people about environmental sanitation. Build a model latrine in each section. And we need to fumigate all the houses. I need everything done as quickly as possible."

"That's a great plan, Dr N_____ and my plans are similar. But I'm a Peace Corps volunteer and I work with SENASA. I don't work for GTZ."

"You're here to work on the GTZ project."

"I'm here to work *with* GTZ on the project, not to work *for* GTZ."

"It's the same thing. Wait until we get there. I'll introduce you to the volunteers and I'll show you how to use the chemicals for the fumigation."

When we pulled into Guazú Cuá, people waited at the *Centro de Salud*, including Gaspar and Lucilla. No other volunteers were there. While the team treated the people who waited with sick children, Dr N_____ showed me the fumigation equipment. The label on the chemical drums had instructions in German and in Japanese.

"I must tell you that I'm not going to use these chemicals. I don't even know what they are."

"They're harmless. They just kill insects, like the *vinchuca*."

"I'm not fumigating houses."

"Okay. I'll speak to your supervisor about it."

"That's a good idea. I'll speak to her, too. As I understand it, you have a contract here with GTZ. You seem to expect me, a volunteer, to complete your tasks."

"You don't understand. I'll speak to your supervisor and she can explain it to you."

"That's fine, Dr N_____. I'll wait. And now I'm going to get something to eat."

Dr N_____ left without looking for me. I saw the white Mercedes Benz bouncing along the road toward Chircal and Escobar.

Later that Friday, May 12, my site coordinator, Vita, arrived. I told her the story about Dr N_____.

"I'll speak to him," she said.

"I'm paying 35,000 guaraníes a month to eat with the Gonzalez family."

"That's way too much. Why don't you cook for yourself until you can find a good price?"

"I plan to. And Gaspar, from the *Centro de Salud*, said his sister would do my laundry for 450 guaraníes for a dozen pieces."

"It should be around 120 guaraníes. Look around. Let people know that you know what things cost. I'm going to visit some families to tell them about you and about Peace Corps. I'll be back on May 30 to formally introduce you to the community. It's your job to invite people."

"Okay, Vita. Thanks. I'll see you then."

I watched her drive her green Toyota pickup across the pasture to the road and disappear into the woods.

9

I got the pictures from my first visit that I had developed and walked around giving them to the people who were at the baptism in May. Everyone wanted to see all the pictures. Everyone offered to pay for the pictures but I refused money. Noemi gave me a sack of grapefruit for hers. The next morning I found another sack of grapefruit outside my door, an anonymous payment for pictures. I received peanuts and *mandioca*, too.

I finished and seeded the soil of two *tablón*es on a warm, clear day with a wind from the east. I planted garlic, onions, parsley, carrots, green chilli, bell peppers, and tomatoes. I planned one more *tablón* and then I'd work on a fence. The *Centro de Salud* had a barbed wire fence, but I needed a better fence around the garden before the sprouts pushed up.

I hauled water from the well to my bathroom and took a bucket bath. I sat in my room and listened to the Voice of America on my short wave radio. Two candles burnt on the table. Clothes hung from both doors of my open *ropero*. Books were scattered about and my mosquito net covered the bed. I couldn't read so I crawled beneath the mosquito net and slept.

I awoke to a steady rain and looked out the window at my garden. Chickens scratched and pecked for seeds on my muddy, destroyed *tablón*es.

I went to Antonio's and he invited me to eat tortillas and

mandioca with him. Noemi stopped in to buy flour while I sat eating.

I saw her again later at the Gonzalez house grinding corn. I asked Gaspar about her and he said she was a single mother and about twenty-four-years-old. She appeared to be different than the other Paraguayan women I had met in that she seemed more independent and confident.

The Gonzalez family consisted of fifteen children. Only the twins Ada and Eda and their older sister Sofía lived at home with their mother, *Ña* Amalia. *Don* Gonzalez and Sofía's husband both worked on an *estancia* and only came home on weekends. The others had moved recently to Ciudad del Este on the Brazilian border.

Rain fell for another day and then it stopped. The sky cleared. The night vibrated with constellations.

I rebuilt my *tablon*es and finished the third and then I grabbed a machete and went to the woods to cut branches and vines for my garden fence. I cut four corner posts first, dragged them to the *Centro de Salud*, and set them in place.

I went back to the woods to continue cutting. Two teenagers came along on their way home from the fields. One held what appeared to be a bloody macaw. It was alive but they had shot it just above the right leg.

"Do you want it?" one asked.

"No. What are you going to do with it?"

"Just show it around, unless someone wants it."

They went on their way and I wished I'd taken the bird. It may have lived.

Guazú Cuá means the place of the deer in Guaraní. There must have been plenty of deer at one time, but no sane deer would come near the place. It would be shot and eaten.

I cut and hauled my posts from the woods to the garden. Noemi's father, *Don* Icho, rode up on his horse.

"Why don't you cut what you need and then come get me? I'll haul all you cut in my ox cart."

"Thanks. I appreciate that."

I called on *Don* Icho after I'd cut a pile of branches and vines. He lashed the horns of his oxen to their yoke and I rode with him in the ox cart to the woods. I loaded my wood onto the cart and we hauled it to the *Centro de Salud* and dumped it near the garden-to-be. *Don* Icho invited me his house to eat. I washed and grasped a hoe I'd borrowed from him and walked to his house. Noemi brought out two chairs. I sat and waited for *Don* Icho to finish with his oxen.

He joined me and talked about his time in the Chaco War from 1932 to 1935. He was a thin man with no teeth but a full head of silver hair. He didn't appear to be old enough to have fought in the war, but he said he was born in 1914.

I remembered reading a short story, "El Pozo," by Paraguayan writer Gabriel Casaccia about thirsty soldiers digging a well in the Chaco during the war.

Don Icho said he'd fathered twenty-plus children with a few women, including nine children with his current wife, Buenaventura. That meant in age he distanced his youngest son by sixty years or so. His wife could have been his daughter.

Many men of his generation had various partners, possibly as a result of the War of the Triple Alliance. The war raged from 1864 to 1870 with Paraguay fighting against Argentina, Brazil, and Uruguay, the allied countries. Depending on which account one uses, Paraguay had an estimated pre-war population of 525,000 and a post-war population of about 221,000, of which approximately 28,000 were men, or an average of about eight women for every man. In the hardest hit areas of the country it was an estimated twenty to one.

Francisco Solano López, son of a Paraguayan dictator, declared war against Brazil. López took power of Paraguay in 1862 upon the death of his father, Carlos Antonio López (1792—1862). He ruled with his mate, Eliza Lynch (Madame Lynch), an Irish woman he met in France, and with whom he fathered at least six children.

When Argentina refused to allow López to march troops through her territory, López also declared war against Argentina. Uruguay joined in an alliance with Brazil and Argentina in May of 1865. After this, Paraguay fought defensively for the duration of the war.

López fought his final battle and died at Cerro Corá (now a national park) on the Rio Aquidabán in northeastern Paraguay. His oldest son was also killed. Brazilian soldiers took Madame Lynch as a prisoner and she eventually returned to France where she died in 1886.

The subsequent provisional government ceded about 55,000 square miles of land to Brazil and Argentina. Brazil occupied Paraguay until 1876.

***

"Come to the table," Noemi called.

We sat at a table on a dirt floor under a straw roof. *Don* Icho and I ate and Noemi served us. My place at the table had been prepared with an embroidered cloth on which Noemi placed my meal in a plate different than that of her father's.

"Why do I have the special setting?"
"You're the guest."
"But I can put my plate on the bare table, too."

"No. It's a sign of respect for you."

"Doesn't your father get the same respect?"

"Of course, but he lives here."

"I want to learn to live as you do. How can I do that if you treat me differently?"

"My mother would never allow us to treat you any differently than as a special guest."

We ate polenta with fried eggs on top and *mandioca* on the side. Noemi squeezed lemons and served lemonade.

I thanked the family and excused myself. I stopped by the Gonzalez home to tell *Ña* Amalia that I'd already eaten. My time would soon be up with them. I returned to the *Centro de Salud*.

As I tried to sleep I thought of the families I knew. They lived without indoor plumbing or electricity. They cooked on open fires on dirt floors and entire families often slept in one room with a straw roof and walls of rough-hewn vertical planks with space enough between them to look out at the stars at night. Feet, horses, and ox-drawn carts were the common forms of transportation. Women drew water from open wells and washed clothes by hand. And these families seemed content.

My purpose involved teaching people how to protect water sources, build sanitary latrines, and to combat intestinal parasites, among other things. Although these were all beneficent ideas, I suspected that the people of the settlement of Guazú Cuá had something to teach me.

In the morning I took my bucket to haul water from the well. As I stepped outside I saw a massive brown pig asleep and sprawled across my destroyed *tablon*es. I pelted the swine with chunks of broken bricks prompting it to gravitate snorting to more amiable grounds.

I spent the morning setting posts and tying vertical-standing branches together with vines. I left one part low enough so I could step over it instead of building a gate. As I sweated, occasional passersby stopped to greet me. They spoke among themselves in Guaraní and I imagined them commenting on how quickly I'd mastered local construction and agricultural methods.

At noon I washed up and went to the Gonzalez house for lunch. Ada served me a bowl of beans in which I found three worms and I wondered how many I failed to find. The girls found my aversion to worms in my beans laughable. Ada explained that the worms were in the beans and couldn't be seen until the beans were cooked, but a few worms wouldn't hurt me. I didn't use their sugar because I could see the ants in it. I didn't know about the beans.

I left a bit nauseated, yet I felt I had made progress. There was a time when I would not have taken another bite if I detected a worm in my food. Today I pushed the worms over to the side of the bowl and continued eating. It went along with the cold splash baths, outhouses, and the semi-potable water.

When I rebuilt and seeded my model garden for the third time, I looked upon my work and imagined plants heavy with juicy tomatoes, fresh garlic sprouts, silvery-green broccoli, bell peppers and romaine lettuce—an area crowded with flowering vines climbing the fence. I'd give away most of the harvest.

I gathered my new map and a notebook and left the *Centro de Salud* to explore Guazú Cuá and to meet the people I'd be working with for the next two years. This would be the beginning of my community census.

I started behind the *Alcaldía*, the mayor's office and, in this case, the police station, near the school and counted my paces to an *almacen* near where I had first entered Guazú

Cuá, noting the paces at each fence corner along the way. I wanted to confirm my location on the plat. I stopped at the *almacen* to ask for water and a barefoot woman with grey hair tied in a bun, wearing a dress so faded the original pattern couldn't be identified, invited me in to drink *tereré*.

She led me to the back of the house to a shady area and gestured toward a chair. I sat. She entered the house and returned with a *guampa* filled with *yerba mate*, a pitcher of water, and her husband. She introduced him as *Don* Bareiro. He grabbed my hand and squeezed it like the grip of a vise.

Don Bareiro sat in a chair near me and *Ña* Bareiro stood and poured water into the *guampa* and passed it to her husband. He sucked the liquid through the *bombilla* and passed it back to her. She refilled and passed it to me.

Don Bareiro stared at me through bloodshot dark eyes with a smile on his brown face topped with uncombed snow white hair. I gave the *guampa* to his wife and she continued to serve the *tereré*.

"Are you looking for a Paraguayan girl?"

"No. No, I'm here with Peace Corps. I'm going to work with the community."

"Paraguayan women are the best."

"They are pretty, I've noticed, but that's not why I'm here."

"The young ones are the best."

"I'm actually conducting a census."

"We can introduce you to some nice Paraguayan girls. Young ones."

"Thanks, *Don* Bareiro, but that's really not why I'm here."

"You don't like women?"

"Oh, yes, very much, but ... "

"Then what's the problem?"

"It's that I'm working or trying to work and get to know the people."

"Then get to know the women. They're the best. Look at this one," he gestured toward his wife. "She's had twenty children that lived, and a few that died."

"Twenty?"

"Twenty," *Ña* Bareiro repeated, smiling.

"Juanita!" *Don* Bareiro yelled.

"Our youngest daughter," *Ña* Bareiro said.

A slender woman with short brown hair came around the corner of the house drying her hands on her plaid skirt that had seen better days as a uniform. She stood by her mother.

"This is our youngest, Juanita. She'll be twenty-one and she's still single."

She smiled, showing a silver tooth, as she glanced at me and reached out her calloused hand. I shook her hand and she said she had to get back to washing clothes and disappeared around the corner of the house.

"She's a very hard worker."

"I can see that."

Ña Bareiro handed me the *guampa*. I drank and gave it back to her.

"Thank you. I'd better get back to work."

"Come by anytime."

"I will."

"When will you come?"

"Soon. I'll come to get information from you for my census."

"Tell us before you come and you can eat with us."

"Thanks, I will. *Chau*."

"*Chau*."

I walked out to the road and opened my map. Right there, catty-corner from the Bareiro house, I saw a one-room, straw-roofed house made of rough-cut planks. Trees and brush surrounded it, so I hadn't seen it as I walked up. I wanted to find a place to rent eventually. I didn't want to spend two years living in a community building or with a family. Perhaps I could rent this house. But I'd finish the census and get to know the area first, I thought.

I knew my location on the plat and decided to start at the *Centro de Salud* again. I'd visit each residence, ask my census questions, and I'd draw the position of each house on the plat.

I stopped at an *almacen* I hadn't been to yet about 200 metres from Antonio's to see what merchandise it offered. An old, stooped woman with a different sandal on each foot and her cheeks sunken in on bare gums met me at the gate. She led me to the window to view the items for sale: a few cans, a couple of packs of cigarettes, half a dozen litres of wine, hard rolls, and gum. I said I just wanted to visit because I would be conducting a community census and the woman nodded her head of thin white hair and understood not a word. She called a name and Noemi and another woman her age appeared around the corner of the house, both barefoot and in tee-shirts and shorts. They'd been washing clothes together. They explained to the woman in Guaraní what I was doing, but she just nodded and disappeared into the house. Noemi invited me to eat at her house that afternoon and I accepted. I hadn't found the food at her house repulsive and I hadn't found any worms.

I left the *almacen* and stopped at the Gonzalez house and completed my first census. I knew the house and the family, so it was the easiest place to start.

I recorded the size of the house, the construction materials, type of floor, cooking arrangements, water source, the outhouse and condition, and the names and ages of the people living in the house. I left feeling accomplished, knowing I'd completed about one eighty-fifth of my census.

I set off for another house I'd passed earlier. I clapped at the fence and a thin man with a wide-brimmed straw hat and a length of rope knotted at his waist as a belt stepped out of the house and waved me on. I ducked between strands of barbed wire to enter the property and met him in front of the house. He'd put out a couple of chairs on the shady side of the house.

"Welcome. Thank you for the visit," he said, holding out his hand. "Toribio."

"Marcos," I said, shaking his hand.

"Sit, sit."

"Thanks. I'm here with Peace Corps."

"I know."

"You do? Okay, well, I'm conducting a community census and I'm going to visit all the houses in Guazú Cuá."

"Good, good. Is there anything I can do to help you?"

"Yes, thanks. For now you can help me with the census of your house."

"Do you drink *tereré*?"

"Yes, I do, but I've had enough for now, thanks."

"What is this census?"

"I'm gathering information on each house in the community so I can determine needs and make a plan for my project in environmental sanitation."

"You were at my parent's house this morning."

"I was? You're one of the twenty?"

"They told me. Yes, I'm one of the younger ones. They want to invite you to dinner Saturday."

"*Don* Bareiro? What time?"

"In the evening."

"What time in the evening?"
"Early."
"Early like 7:00?"
"About then."
"I'll go. Now, may I ask you the questions for the census?"

I jotted down a description of the house: vertical-standing planks as worn as driftwood with a thick straw roof that had greyed from age, and a dirt floor in the one room. A lean-to open on one side served as the kitchen. A three-sided coconut bark roofless structure allowed privacy for those bathing or responding to nature's call.

We finished the census and Toribio offered to take me to his brother's house and a few other houses. Toribio cultivated cotton for cash and some corn, beans, and peanuts for family use. He had a field slashed out of the woods a couple of kilometres from his house.

I ended the day with census information from five houses and I determined the position of each house by pacing along fence lines from known points. I measured the perimeter of the houses by pacing, too, and the distances to outhouses and water sources.

I walked back to the *Centro de Salud* to get cleaned up before dinner. I realized that people wanted me to visit, they wanted to know me. Not many strangers passed through Guazú Cuá and I believe I was the first person from the United States.

At Noemi's I sat with *Don* Icho, *Ña* Buena, Noemi and two of her brothers, Joel, about five and thin, and Rodolfo, about eighteen. We ate a stew of pasta and meat with carrots with *mandioca* on the side. After dinner, Noemi helped me with the census information for the house and the family. Rodolfo mounted a horse and rode off to herd his cattle and sheep into the yard for the night. I measured the house.

Three houses stood side by side. On the fence side, facing the *capilla*, was a two-room brick structure with a tile roof. It butted against a one-room wood house with a straw roof. Parallel to that structure, two metres distant, was a wood and straw building with one room for sleeping and a semi-closed cooking area on one side and an open area covered by a straw roof where meals were served.

A well, about three metres deep, had been dug and lined with brick behind the middle building. A coconut bark conduit channelled rain water from the roof to the well. Three metres from the well were the outhouse and *bañadera*, a bathing stall, brick with a tile roof. The outhouse had a clay pot as a stool and a user would pour water into the pot to wash the waste into a cesspool beside the outhouse. Neither stall had a door, but ragged cloth for privacy.

10

Some days later, my Site Coordinator, Vita, and Luci from *Proyecto Paraguarí,* lugged into Guazú Cuá in a Peace Corps pickup truck. We met at the *Centro de Salud* to wait for the population to arrive for my formal introduction. The three teachers, Vilma, Andresa, and Nidia came. So did the Gonzales family, Antonio, Toribio, and the police chief. Vilma, the school director, told the students to invite their parents, but she gave them the wrong date. Vita introduced me to the people I already knew.

I decided it would be a good time to go to Asunción to collect my mail and my living stipend, so I rode out with Vita and Luci.

Vita dropped off Luci in Paraguarí and me at the Peace Corps office in Asunción. I collected my mail and caught a bus downtown.

I ran into several members of Areguá-1 at a pub near the Stella Hotel, where many volunteers stayed. Some had been in town for a week already for no apparent reason other than boyfriends or girlfriends or appointments with the Peace Corps doctor. Others waited to leave.

One woman told me she'd been the victim of an attempted rape. She had been walking to her house with some bags and a man on a motorcycle from her site stopped and asked her if she wanted a ride. She said she didn't, so he offered to take her bags for her. He said he'd drop them off at her front gate. She agreed and he left with the bags and she continued walking.

When she reached her house, the bags waited at the front gate. She carried them inside her house, put her things away, and then went outside to use her outhouse. Half way to her outhouse the man that delivered her packages jumped from hiding and attacked her. She fought and screamed and ran. Neighbours came to her aid. She had been on the bus that crashed during training. The attack at her site brought her to Asunción to request an early termination. She stayed at the hotel for a few days before her flight left.

Another woman from the group had already left. She left a note for the rest of the group. She, too, had been on the bus and had broken her clavicle in five places and wanted a second opinion from her doctor in the United States. She took an early termination.

An older couple in Areguá-1 had served in Peace Corps somewhere in Africa in the 1960s. They hadn't learnt enough Spanish between them during training and terminated early and left for home.

Most of the volunteers I saw were in Asunción to get their stipends and mail, but many had only spent a few days at their sites before returning to take rooms in hotels, as if they couldn't let go of one another, of the familiar, to embark upon a journey to the strange and unfamiliar.

Areguá-1 had lost six members in the first four months.

I took a room at the *Residencial Itapúa* and had dinner with a fellow volunteer at a German restaurant. After dinner we returned to the pub, which, by then, teemed with Peace Corps volunteers from our group and several other groups.

In the morning I arranged to have a reel to reel projector and a generator hauled to my site in two weeks. I rented a couple of films from a business and fitted them in my backpack, surprised at the weight. I bought a few items to do my own cooking and had lunch with a few members of my group. I caught a bus to the terminal and then another bus to Paraguarí. From there I squeezed onto a bus to Sapucaí and got off at the road to Guazú Cuá after dusk at 7:30 on this late autumn evening with fifteen kilometres to go.

I adjusted the thirty-five-pound pack on my back and started walking along a disappearing road. Clouds obscured the stars and no moonlight reflected on the ground. A strong north wind blew. I fell in a deep ox cart rut in the road and I tripped and fell to my knees a score of times before I stopped and looked back at the total blackness. Ahead of me were hills to climb and creeks to ford and darkness. I listened to an owl hooting and something scrambling through the woods beside the road. I had walked into the darkness so deeply that to turn back would have been the same as going forward. I continued and spotted the occasional flicker of a candle or a fire in a distant house. Wind rustled leaves and stands of bamboo. Animals snorted in the night. I smelt fresh manure as I stepped in it and I tripped hundreds of times before I reached the *Centro de Salud* three hours after getting off the bus. No local would have walked in alone in the blackness, but I knew no better.

My time ended with *Ña* Gonzalez and I told her I planned to prepare my own meals. She displayed her dismay at the loss of income by ignoring me. The girls begged me to stay. They suspected I'd found another place to eat for a better price and tried to strike a new deal. They liked having the only house in Guazú Cuá that an American visited every day. They invited friends to stop by to watch me eat, as if I were a unicorn or a winged man.

I bought a *bracero,* a one-burner charcoal stove. It held a couple of bowls full of charcoal and the cooking pan sat directly on the coals.

I set up my cooking area outside on the landing of the *Centro de Salud*. I gathered a pile of twigs to get a fire started as I'd seen the Paraguayans do. I kindled a fire and placed a chunk of charcoal on it, which extinguished the fire. I tried again and added sticks until I had a decent flame and added small chunks of charcoal. I fanned the fire with a pot lid and put the fire out. I walked to Antonio's and bought a litre bottle of kerosene and carried it to the *Centro de Salud*. I piled my kindling in the *brasero* and sprinkled it with kerosene, and then I added sufficient charcoal and wet it with the flammable liquid. I corked the bottle and placed it out of the way, struck a match and dropped it on my fuel. Within fifteen minutes I fanned white hot coals.

I fried potatoes and eggs and ate them with hard rolls from a metal plate. While I ate I heated water for coffee. After breakfast I hauled a bucket of water from the well to wash dishes. The grease wouldn't wash off the pan or dish with the cold water. By that time my coals had become hot ash. I built another fire with the help of kerosene and heated water to wash my utensils.

I passed half the morning building a fire, preparing breakfast, eating, and cleaning up afterward. I figured I'd better skip lunch and return early enough to repeat the process for dinner. I wouldn't get much done if I spent two years gathering twigs to build cooking fires.

That evening I stopped by the Medina home to visit and *Ña* Buena invited me to stay to eat. I accepted. I ate with *Don* Icho and Rodolfo. Noemi had gone to Asunción to help her sister for several days. *Ña* Buena served *mbejú*, a pancake-like mass prepared with *queso Paraguayo* (homemade Paraguayan cheese), corn flour, and *mandioca* flour. It's often served with a fried egg and with milk to drink. I hinted that I may want to find a family to eat with since my obligation to the Gonzalez family ended, but let it go. I thanked *Ña* Buena for the meal and walked back to the *Centro de Salud* full and happy, knowing that I wouldn't be hauling water and dousing charcoal with kerosene that night.

***

One morning I sat outside in the chilly air, unkempt, somewhat frustrated, preparing my breakfast when Dr N_____ and his entourage arrived.

"Good morning, Marcos."

"Good morning, Dr N_____. What brings you here on a Thursday?"

"Friday."

"Today's Friday?"

"Yes, it is. What have you accomplished since we last met, Marcos?"

"I've been working on a community census and plotting the locations of houses and water sources on a copy of a plat, which I'll transfer to my own map on Mylar. I'm going to have a movie night to raise funds. And I finished my garden." I gestured toward my rickety fence behind the building.

Dr N_____ chuckled at my garden. "With the exception of the garden, you haven't done anything for the project. A census is great, but I need these houses fumigated."

"With all respect, Dr N_____, I explained that I wouldn't use the chemicals, and that I'm a Peace Corps volunteer. I don't believe I'm obligated to complete a project that you are apparently being paid to do. I'm here for the community and I will be happy to help you with your project, but as a volunteer."

"Okay. I'll speak to the Peace Corps director. If you don't want to work, you shouldn't be living in the *Centro de Salud*, because this building is part of my project."

"I thought a foreign entity built it for the community and not necessarily for your project?"

"It's my responsibility now, and I requested a volunteer to work on my project."

"But usually the community requests a volunteer, Dr N_____, not a contractor. You might have been better off hiring someone. I'm a volunteer, not free labour."

"Leave it at that, Marcos. I'll get back to you through your Peace Corps director. Maybe you shouldn't be here."

"I believe I should be here, but I'll wait for word from Peace Corps. Until then, Dr N_____."

***

The doctor left me on the landing with my breakfast. His attempt at intimidation got my ire up. I put my dirty utensils in my room and left the building for the day to continue with my home visits and census.

I walked to the house which stood in a grove of trees to check its condition. The straw roof had a newish golden colour and a fresh smell to it. Rough cut, unpainted, vertical-standing planks formed the walls with no battens to close the spaces between the planks. The one dirt-floored windowless room had space enough for a bed and *ropero* and a table. The roof extended over the entry and formed a verandah the same size as the room. I found no outhouse or water source.

I left the dwelling and started up a hill towards the owner's house, another Medina family. The house sat at the top of the hill at the edge of the woods and faced an open pasture. I thought I'd have to put in a brick floor, and build a latrine, a bathing stall, a kitchen area, and close the spaces between the wall planks to keep out the wind and rain.

Don Medina met me at the door and invited me to sit. He called his wife to prepare *tereré*.

"You can live there if you're going to improve it and take care of it."

"I'll let you know in a couple of days. I saw another place I want to look at first."

"No problem."

"I'm conducting a community census. Do you have time to answer some questions for me?"

"Whatever I can help you with."

I gathered his information and then located his well and house and outhouse by pacing along fence lines and turning right angles to the structures. By the time I finished it was midday and the family invited me to stay for a lunch of stew and *mandioca*.

Saturday evening I went to Toribio's parents' house for the dinner party. As I approached I heard music and laughter. At the house saddled horses stood idly with their reins tied to the fence. Two dozen people sat or stood in front of the house. Four men, three with guitars and one with a harp, played and sang songs in Guaraní. Men passed around glasses of beer and *caña*. A couple danced on the grass. It was Toribio's sister's birthday and I didn't know. Men outnumbered the women. Girls served paper plates of *empanadas* and *mandioca*. I shook hands all around, and noticed some of the young men already had cast off all social inhibitions behind alcohol.

I ate some of the food and then asked the birthday girl to dance. I took her by the hand and led her onto the uneven grass and positioned her to polka. She wrapped her arms around me and smiled with the celestial light glistening off her silver tooth. I dragged her around the yard for a couple of songs to the applause of the guests and then I bowed and led her back to her seat beside her mother. I thanked her and her mother and excused myself. I said I had to get up early for a meeting. I slipped out the gate and past the snorting and stomping horses.

11

I sat on a bench in the breezeway of the *Centro de Salud* the following morning before 8:00. The *Alcalde*, the mayor, had called a meeting to ask for help in painting the *Alcaldía*. Sofía Gonzalez would introduce me to the community.

8:00 passed and 9:00 passed and I thought I'd heard wrong. Yet I waited, and at 9:30 two men walked up to the breezeway where I sat. We shook hands and they confirmed that there was a meeting. An hour later a dozen men had gathered and sat talking and drinking *tereré*. The *Alcalde*, who lived in the *Alcaldia* just fifty metres from the *Centro de Salud*, noticed that enough people had arrived to begin the meeting so he sauntered over in his rumpled fatigues and cap with a pistol strapped to his waist. When the Gonzalez girls saw the *Alcalde* walk toward the *Centro*, they, too, appeared in the breezeway.

The *Alcalde* shook hands all around and announced that he had called the meeting to organize a group to raise funds and volunteers to paint the *Alcadia*. Sofía stood and said that she'd like to introduce the community to its Peace Corps volunteer.

"This is Marcos," she said, gesturing toward me. "He's going to show movies here." She stood back, smiling.

"Movies?"

"When?"

"How?"

"I have the film," I said. "I'll soon have a projector and a generator. Maybe in a week."

"What are you going to do with the money you make?"

I hadn't thought too much about it. "It's community money," I said. "It's for environmental sanitation."

Men talked amongst themselves in Guaraní and gestured toward me. I understood not a word. I stood and walked to the centre of the circle and asked the group to help me.

"I'm here for the good of the community," I said. "But I need your help. My Spanish is basic and my Guaraní is almost non-existent. I must be able to communicate with you before I can work with you. For example, listen to this."

I spoke in English for about thirty seconds. No one understood a word I said.

"That's what Guaraní sounds like to me," I said. "I need your help so that we can communicate."

I looked at the affirmative nods from the crowd. I didn't see Antonio or the Medinas or the Bareiros. I saw mostly unfamiliar faces.

Two men, *Don* Osorio and *Don* Nuñez, invited me to visit their houses at any time and they gave me rough directions, pointing over a hill I hadn't crossed yet. Another, Jorge, invited me to a horse race and said he'd stop by for me after the *siesta*.

Later Jorge clapped his hands outside the *Centro de Salud*. I went out to meet him where he stood at the gate, smiling with curly black hair sticking out from a blue baseball cap. His faded yellow shirt hung over a threadbare pair of slacks that had once been part of a suit. He wore black *zapatillas* on his feet. He held out his brown, solid arm to shake my hand. I shook his calloused hand and we walked across the grassy *campo communal* toward the site of the race.

We sat on the ground beside the track, two parallel dirt trails beaten into the pasture. Men arrived from all directions. Antonio had his ox cart in the shade and he sold beer, *caña*, soda, cigarettes, and candy. The track stretched a 500 metres along the high bank of a small creek. Whitewashed posts marked the starting point and the finish line. Two men rode horses bareback near the starting line.

"What do you do here, Jorge?"

"Anything," he said, rolling a cigarette. "I'm a mason, I work in the fields, I pick cotton—I do what I can." He lit his cigarette and took a long drag. He gestured to me with the fixings and I shook my head.

"You have to be careful here," he said.

"In what way?"

"You're not from here and people don't know why you're here. They don't believe that someone from a rich country would come to live in a place like Guazú Cuá as a volunteer. Some people think you might be a spy and are here to collect taxes."

"I sensed that. Some people want to take advantage of me."

"That's where you have to be careful. Some people will overcharge you because they know you don't know the prices."

Jorge smoked and stretched out on the grass. "I'm not from here, either. I was born in Paraguay but my parents moved to Argentina when I was about five. I lived there for twenty-five years and then ended up here. I'm contraband. I have no papers. I have no land, no horse, and no animals except a few chickens. I have to work for other people and sometimes it's not worth it to work because people don't want to pay."

Jorge sat up and looked toward the starting line. Men tried to position both horses at the starting line.

"Here they go," he said.

Men in broad-brimmed hats walked from group to group

taking bets. Near the starting line I saw someone raise a pistol and point it skyward. Smoke puffed from the barrel and the discharge cracked. The horses bolted from the starting line and their riders reined them in, turned and trotted back to the starting line.

"What happened?"

"They didn't start at the same time," Jorge said.

The race started five times before the horses and riders pounded past us and finished the course. One beat the other by two lengths.

"That's it," Jorge said. "Let's go."

"That's it?"

"There are only two horses today."

We walked back to the *Centro de Salud*.

"Come by my house to eat and talk tomorrow night," Jorge said.

"What time?"

"Early."

"What time is early?"

"About dusk," he said, walking toward the school to his house as the late autumn sun dropped with the temperature.

12

I walked to a part of Guazú Cuá I hadn't been to and stopped at a dwelling. Osorio, the man of the house and a community leader, sopped up egg yolk from his plate with a hard roll. We sat at a table in the shade in front of his house. I saw women in an open storage room in the house scraping dried corn kernels from cobs onto a cow hide. They stacked the cobs against the wall. A corral near the house held two horses and a dozen Brahmas.

"Would you like more to eat?" Osorio said.

"No, thank you."

"Let me see your map."

I unrolled the plat on a cleared area on the table. Osorio studied it.

"Here we are," he said pointing to the map. "This is my twenty-five hectares. I've been here for thirty-two years. I'll show you around."

Osorio put on a straw hat and we walked behind the house where the land sloped toward a creek. Cotton and corn grew on several hectares and the rest of the property remained wooded except for an area littered with felled and blackened tree trunks.

"We're going to haul out that timber so we can plant more cotton," he said. "This was all thick woods when I came here. I cut the trees, burnt the area and pulled the stumps with axes, shovels, and oxen."

We walked through the field to the burnt area. I heard

monkeys chattering in the standing trees beyond the cleared area but I couldn't spot them. Osorio kicked at the ground and reached down for a hand full of black topsoil.

"This is good soil," he said, dropping the dirt. "It's just hard to get to."

Back at the house we drank *tereré* and I sketched the house and outhouse and the water source, a spring that Osorio had tapped. He built a small pool to gather water. I completed my census information and Osorio walked with me half way to the *Centro de Salud*. The *Centro de Salud* sat on the opposite side of the hill. Osorio left me at the hilltop where I had view of the centre of Guazú Cuá.

I continued down the hill with a view of open pasture before me and woods on either side of me. On my right I saw *Don* Medina's house and to my left, through the woods, I saw *Don* Cazares' house. At the bottom of the hill I turned toward the *Alcaldía* and the school and then past the *Centro de Salud* to Antonio's. There I bought tomatoes, carrots, and bell peppers as a gift to *Ña* Buenaventura, who had invited me to lunch.

I clapped at the gate and *Don* Icho invited me to enter. He waited at a table under the *galpón* (a shed or storehouse, often open-sided). I dropped the vegetables on the table. *Ña* Buena thanked me and served us lunch. She seemed to have forgotten that she'd invited me. Or maybe I had misunderstood. At any rate I felt out of place. I ate quickly and took my leave and went to the *Centro de Salud* for a nap. The *siesta* was not a good time to go out visiting houses, anyway.

Clouds built and darkened. I smelt rain but didn't see any. Few things moved during the *siesta*, except for the horses, cows, and sheep grazing in the *campo comunal*. The distant sky blackened but stayed grey over Guazú Cuá. The breeze picked up. I glanced at my barren garden and then slept for an hour.

When I awoke the black clouds had increased and the air felt damp and smelt fresh as the wind gusted briskly. I took my map and walked to *Don* Cazares' house. I slipped through the barbed wire and walked across the pasture below his house. The wood and straw house sat on level ground protected by trees and rows of flowering plants. As I neared the house a boy ran out to lead me the rest of the way.

Don Cazares and his wife and a couple of sons sat outside drinking *tereré* and looking at the approaching storm. A young boy wobbled in a wooden armed-chair. His skinny, twisted legs stuck out from the chair and he held one arm as if he couldn't use it. He smiled with crooked teeth and didn't take his huge black eyes off of me. A few yellow-headed parrots with clipped blue wings watched me from a branch.

"Welcome. Have a seat," *Don* Cazares said, pointing to an empty chair. A broad smile broke behind his bushy, greying moustache. A wide-brimmed, beige hat topped his lanky form. He spoke mostly Guaraní and little Spanish.

I sat and he passed me the *guampa*. I drank and passed it back. He squinted into the *guampa* and tapped it with his knuckles. We had an awkward conversation because of my problems with the language, yet I felt comfortable with *Don* Cazares and his family.

"Julio will be six next month," *Don* Cazares said, looking at the thin, twisted boy. "We would like you to come to his party."

"I will. And I can take pictures," I said. I knew I was probably the only one in the community with a camera.

"How much do you want for each picture?"

"I don't want anything. I'll give them to you."

"We have to give you something."

"No you don't. Maybe invite me to eat one day."

Don Cazares said something in Guaraní and his wife got up and walked behind the house to the cooking area. No one had spoken except me and *Don* Cazares. The others just looked at me. She returned with a sack of grapefruit and sweet potatoes and gave it to me.

"Thank you very much. I'd like to come back to get my census information," I said, standing.

"Anytime. Our house is open to you," *Don* Cazares said, reaching for my hand and gripping it tightly.

"Thank you. I'll return soon."

I walked back to the *Centro de Salud*, dropped off my sack and then hauled a few buckets of water so I could bathe before I went to Jorge's house. I stood in an aluminium basin and poured cups of cold water over myself and then scrubbed myself with soap and lathered my hair. I scooped out cup after cup of water to rinse the soap and shampoo off. I used less than five gallons. I saved the runoff to use in the pour flush toilet. I dressed and started a letter to my sister while I waited for dusk.

I left in the darkness of the incoming storm and the absence of sunlight. I carried a pocket flashlight to see the cow droppings in the road. A gust of wind blew a sheet of horizontal rain into me as I walked. I reached Jorge's one-room wood and straw house wet. He called me from under his *galpón* where he sat smoking in the dark.

I sat at a small table with Jorge. A dim light flickered inside the house and charcoal smoke curled from the room and vanished with the wind. We shook hands. Rain water dripped from the straw roof.

"Do you drink *mate*?"

"Every day."

"Julia, is the water ready?" Jorge said.

"Yes," Julia said stepping through the door with a tea pot in one hand and a gourd and *bombilla* in the other. She pulled up a chair and sat with us and served the *mate*. Her curly hair looked like a brown halo around her head. She wore a blue skirt that passed her knees and that had more patches than original material. The short sleeves on her over-sized tee-shirt passed her elbows.

"This is Julia. Julia, meet Marcos," Jorge said.

"You were at my house today," she said, as rain water puddled in the yard.

"Where is your house?"

"*Don* Cazares is my father, and Julio, the boy you will take pictures of on his birthday, is my son."

"They were both born in July," Jorge said.

"Julio stayed at my parents' tonight because we didn't want to get caught in the rain with him."

"Julia was married and when she was pregnant with Julio her husband punched and kicked her in the stomach. She ran away and walked twenty kilometres during the night to her parents' house. The baby was born, well—you've seen him."

We finished the hot water in the tea pot and Julia went inside the room and returned with a plate of Paraguayan *tortillas* and *mandioca*. Paraguayan *tortillas* are prepared with a batter of wheat flour, *queso paraguayo*, eggs, and milk, and sometimes with green onions and Swiss chard. The batter is fried a scoop at a time.

"Here you go," she said, as she sat again.

"And you?"

"We're okay. Go ahead and eat."

I ate while Jorge talked. I couldn't help thinking that I

was eating the last of their food. Julia watched me eat. The rain slowed to a drizzle and a choir of frogs whined like babies.

"No one owns their land here except one family," Jorge said, grinning. "The government subdivided this area and opened it for settlement. Anyone could settle on land and pay for it later. That was years ago. No one ever came to collect," he laughed. "The government never finished the access road. I claimed this lot and built this house and dug a well. I don't own the land, the government does."

"What if someone came to collect?"

"There are about eighty families here. If we can't pay, what will the government do with us? Only one family paid," he said, holding up one finger, "and has a title to its land. They live across from *Don* Icho's, where you like to go."

"The *almacen* across from Noemi's?"

"Yes, but it's more of a place to drink than to buy food."

"Why was that family the only one?"

"People formed a cooperative at the house. *Don* Gonzalez thought if he owned the cooperative, the building, at least, he'd profit. He went to Asunción, paid, got his title to the land and then the cooperative broke up because no one trusted him anymore."

"But plenty of people still go there."

"To drink. Not to have a cooperative," he smiled.

"I think I understand." I watched water drip from the straw.

"What about you? Why did you leave your home to come here?"

"I'd always lived with indoor plumbing, electricity, transportation, entertainment, and with access to whatever I needed, Jorge, even though I lived in a low income bracket."

"We have none of that here in Guazú Cuá."

"I know. I knew that most people on earth did not have the opportunities that I had, that most people lived differently. I wanted to experience what life was like for most people, and I wanted to help people, too. Peace Corps has given me the opportunity with its three goals."

"You wanted to live like us even though you had the very things we need?" Jorge laughed.

"Yes. I wanted the experience. I knew, but I only knew from reading. I wanted to know how so many people in this world live."

"What are these goals you mentioned?"

I finished eating and wiped the grease from my hands on the tablecloth.

"Thank you for the great meal," I said. I continued, "One is to help people in places like this with trained personnel. Next is to help you understand my culture. Finally, to help Americans like me to understand other cultures such as yours."

"The people here don't want to change."

"They don't have to change. But they might want to protect water sources and build sanitary latrines. And improve the school. And build *fogones* instead of cooking on the floor."

"That's all change, but I'll work with you, Marcos. Don't expect miracles."

"I appreciate that, Jorge. I may need you to help me rebuild Medina's *casita* if I rent it."

"Save yourself the trouble. There are plenty of places to stay without spending money on that dump."

"I want to leave the *Centro de Salud*. I don't feel comfortable there, and Dr N_____ wants me out. I want to live like you live and like the others live. Do you understand?"

"No, but, yes. You're new here. Wait until you know

more about Guazú Cuá and then find a new place to live. Or find a family to live with."

"I've thought about that, too."

"We've noticed you spend time at *Don* Icho's." They both stared at me.

"That could be a possibility, but ... "

"But you haven't been invited?"

"I haven't, and I'm not going to invite myself."

"There are plenty of families you can live with. We'd take you but we only have one room."

"I need a room."

"We'll talk more about it."

I stood and shook hands with Julia and Jorge. "I'd better get back while the rain has stopped. I appreciate your time."

"No problem. We'll help you get out of the *Centro de Salud*."

I sloshed through the puddles and mud to the gate and turned and yelled, "*Chau*."

"*Nos vemos*."

I rolled up my pants and splashed home to the *Centro de Salud*.

13

Back in Guazú Cuá after another trip to Asunción, I stared out my bedroom window where a full moon lit a clear and cool night showing too many plants in my garden. Too many books and magazines littered my table as I sat in the weak light of a candle. I'd walked here last night, three hours from the bus stop. I had travelled to Asunción only to spend money and to be sick.

Some days earlier I had walked out of Guazú Cuá on a Tuesday in June, the day the people from GTZ were to visit. I waited past their scheduled time and then decided to wait for the *mixto* (a truck that hauls all—people, animals, produce), which didn't come either.

I shouldered my pack, grabbed three raw carrots, and off I went on the fifteen kilometre hike to the bus stop. Halfway to the road, near the community of Chircal, I met the GTZ personnel entering the site in a white, Mercedes Benz SUV. The director told me to wait for them at the road and they'd pick me up on the way out and then would give me a ride to Asunción.

I hiked the rest of the way to the main road and rested in the shade beside a creek until I heard the Mercedes Benz manoeuvring its way along the road and around the deep ruts cut by wheels from ox-drawn carts. I walked up the creek bank and stood by the wooden bridge and watched the mud-splattered vehicle approach. It stopped and I got in the back seat. A German doctor drove and when we turned onto the main road, he drove as if he were on the autobahn, leaving a cloud of red dust behind us. He turned right on the main road instead of left, and, as he did, he said he wanted to stop first at Sapucaí to see the machine shop for the train.

Paraguay had the first train to run in South America and the country still ran a steam-fired engine fed by wood. The machine shop had been operating beside the tracks in Sapucaí since Paraguay built its railroad through the area. Carlos Antonio López, the president of the republic, contracted British engineers to build the railroad. Construction began in the mid-1850s.

We sped into the town of Sapucaí, past shops where men constructed ox cart wheels and women led donkeys laden with goods to the market. At the Sapucaí train station we acquired permission to tour the machine shop. We walked through the building as steam-powered machines built in Britain in the late 1800s hissed and whirred and clunked. Men worked without safety gear at grinders, drills, and saws, while others tossed chunks of wood into the fires from carloads of wood that were parked outside. We spent a half an hour in the sooty, smoky, steamy building and then we left Sapucaí as if we were in a race. Dust billowed behind us for thirty kilometres until we reached the asphalt of Route 1, where we barrelled toward Asunción.

The group from GTZ dropped me off at the *Residencial Itapúa*, off of *Avenida Estados Unidos*, a main avenue on a hill almost in the downtown area. A half block away from the avenue, with its constant din of honking horns and diesel engines, sat the guest house, a three-storey private residence. The owners had a dozen rooms to rent to travellers. From there I could walk anywhere downtown near the river and to the *Mercado Cuatro*, a sprawling, busy open-air market where one could buy fresh produce, clothing, electronics, car parts, medicine, medicinal herbs, raw leather, and most other things legal and illegal.

I took a room and doubled over with stomach cramps. I stumbled into the bathroom and barely got my pants down before the diarrhoea flowed. After several trips to the bathroom, I collapsed exhausted on the bed and slept until the next morning.

I still didn't feel well, but I went out to run errands, which included a trip to the Peace Corps office for my mail. I met two women and a man from my group at the office and I went out to lunch with them. I enjoyed speaking English again, but was surprised to learn that they had been in Asunción most of the time and not at their sites. It was mid-June. I'd been at my site for several weeks already. After lunch it was over for me. I left the group and went back to the *Residencial Itapúa*. I spent most of the afternoon in the bathroom with diarrhoea. I couldn't stop shaking I felt so cold, and then I soaked the mattress with sweat as my fever broke.

In the morning I went straight to the Peace Corps doctor and got medication and gave stool samples. The results would be in the following day, Friday. I went back to my room and did nothing, drained. The next day I visited the doctor again. I had an intestinal bacterial infection, probably from the water. This would the first of many intestinal torments, which, if nothing else, kept me lean.

I recovered quickly and planned a trip to Paraguarí for the following Friday.

Rain storms hit every day and left the ground saturated. Water flowed along the surface to brown boiling creeks that ran bank to bank. The doctor wouldn't come because of the mud and the torrents cutting across the roads. I paced the length of the *Centro de Salud*. I wrote letters, read, plotted houses on my map, planned to build a solar oven, sloshed to Antonio's a few times a day, and I wanted to get out of Guazú Cuá badly. Sofía told me that I could walk to the *estancia* (ranch) where her husband worked—an hour's walk—and wait for a truck going to Paraguarí.

"The road from the *estancia* is good and there are always trucks. Any truck will take you. You won't even wait an hour," Sofía said. "That's how most of us get out of here."

I wore shorts and *zapatillas* when I set out early in the morning on an empty stomach, planning to eat at a restaurant in Paraguarí. I trudged across pasture, climbed through barbed-wire fences, waded through water and mud, forded swift streams, and watched the black sky, the cattle, and a few rheas (an ostrich-like bird) on my sojourn to the *estancia*. I reached its fence in an hour or so, but the walk to the buildings and the road took another hour.

I stripped and washed in a rivulet that ran through the brush. I changed out of my dripping, muddy clothes and put on my shoes. I walked out to the road near the area where tanker trucks filled with milk before heading toward Paraguarí and beyond. My stomach growled.

Sitting beneath an august mango tree beside the road, I thought that I might have been in Paraguarí already if I had taken my usual route, but, instead, I listened to falling rain and my empty stomach for four hours. Why did I listen to Sofía? She's probably never been out of town. It wasn't her fault, but mine for taking her advice without a second opinion.

In the distance came the sound of a lumbering diesel, and I soon saw that it hauled feed to the *estancia*. But there appeared to be a very attractive woman reclining atop the load of feed, and as the truck neared and stopped, I saw that it was Noemi. She clambered down, her black ponytail swaying, and went to the passenger-side door where she got her baby, Augusto. I greeted her and she acknowledged my presence and asked where I was going and how I'd gotten to the *estancia*. A pickup pulled up that would take people toward Guazú Cuá and she scrambled into the bed with Augusto. Several others sat in the bed of the truck. It headed toward the last fence I'd climbed through.

I waited for a truck going in the opposite direction and thought about Noemi, her honey-coloured skin, her dark eyes, and her lithe form. I realized I'd been smitten by her since I first met her. It may have been her healthy smile, with no makeup or other adornments, or the swing of her hips as she padded barefoot in a threadbare skirt, tossing her cascading black crown over her shoulders. It may have been her natural contentedness with her lot in life and her friendliness. I found nothing plastic in the woman.

An hour later I caught a ride with the same feed truck on which Noemi had come, and I rode in the cab to Paraguarí, arriving at about 4:00 p.m. just as the early winter darkness began to descend. I realized that Noemi often crossed my mind as we bounced toward Paraguarí.

I ate at the restaurant, *La Curva* (the curve), on Route 1, and thought that Paraguayans often make this trip on empty stomachs and don't run to the nearest restaurant. Few can afford such luxury.

I had several letters to mail, so I left *La Curva* and trotted to the post office and found it open. The smallest bill I had was 5,000 guaraníes. The clerk couldn't change it and she said she'd close in a few minutes. I burst from the post office and ran from store to store asking for change. I couldn't change the bill unless I bought something, so I got some gum. I ran back to the post office and the clerk said she couldn't post the envelopes because they had tape on them, which, she said, was prohibited. I snatched the envelopes and ripped the tape off. She took them.

Next I scrambled around the *centro* of Paraguarí to purchase the items on my list before the stores closed, but without success, so I set out to find a place to spend the night.

The only pension in town wanted to charge 5,000 guaraníes, but I responded as if I knew it should be less and the owner lowered the price to 4,000 guaraníes, so I took it not because it was a good deal, but because I was tired and dirty and I wanted a real shower.

At dawn I was out and about the market purchasing utensils I thought I might need and groceries. A bus for Escobar left at 7:00 a.m. and another at 10:00 a.m. I had to catch the early bus to make it to a lunch engagement at the Nuñez house in Guazú Cuá. I'd have to walk fifteen kilometres from the bus stop. I bought what I could, including aluminium foil and a sheet of glass for a solar oven that I planned to make, but slashed many items from my list.

The bus arrived at the terminal with people on the bumper, hanging from the doors, and holding on to seats inside while they balanced on wheel wells. Before the bus stopped passengers leapt from the bus and others tried to board. I pushed my way on with my backpack and my sheet of glass. Humanity pressed against me on all sides. Hobbled chickens flopped at my feet. I tried to warn people about my glass. A woman by the window held out her arms and I passed my glass to her. Passengers had chickens, piglets, sacks of bread, and clothing, but no one else had a pane of glass wrapped in newspaper.

I wormed my way off the bus at my stop and the woman with the glass handed it to me from the window only missing a chip off a corner.

Backpack in place and glass pane under my arm, I set out for Guazú Cuá and hiked by pasture, cotton fields, up, over, and down hills, past one-room, straw-roofed houses for two, and a half hour to the open *campo comunal* of Guazú Cuá. I saw Noemi washing clothes at her neighbour's house and said hello in passing but continued on to the *Centro de Salud*, which I reached at 10:30.

I heated water, took a bath, dressed, and then walked a few more kilometres to the Nuñez house.

14

Each day at my site, shortened and cool winds blew from the south. With no heat inside any of the houses, the temperature varied little whether inside or outside. On the first evening of winter I wore a thick, wool poncho and sat on a bench in the breezeway of the *Centro de Salud* drinking *mate* and staring at the southern night sky, resplendent with stars and satellites. Thunder boomed in the distance. A storm had recently passed and left puddles and mud, plenty of mud.

Before the storms, I had made a trip to Paraguarí to pick up a solar-powered slide projector, a giant toothbrush, an enormous set of plastic teeth, and a supply of fluoride. I also had a script about a girl who didn't take care of her teeth, and I purchased seventy-two toothbrushes to give or to sell at cost to the students. As I sat outside looking at the stars and sipping hot *mate*, I planned my first day working at the school, which would be the next day, Thursday, if it didn't rain again. Many students walked several kilometres to school, so school often cancelled when the roads became slippery with deep bogs of mud and runoff.

The school consisted of a row of three classrooms with a teacher for each. One group of students attended school in the morning and another in the afternoon. I would visit six classes with my dental health lesson.

I showed up at the school and entered the first class like the main attraction in a freak show. Andresa, a stocky teacher, introduced me as I set up my slide projector. The windows, or the wooden shutters, on both sides of the room were open to allow light in, but the cold breeze from the south passed through as well. About twenty children watched and whispered and giggled as they swung their legs and mostly bare feet. Most had never seen a slide projector, or an American for that matter.

Andresa turned the class over to me and she took a seat by the window, the daylight glistening off her glasses. I reintroduced myself and began my slide show on dental health care with graphic examples of rotting teeth, explaining each slide in my halting Spanish. Andresa hushed the students when they laughed, and I wondered if they laughed at me, my Spanish, or the slides. I checked to see if my zipper was down or if I had part of my breakfast on my jacket.

I finished the slides and told the story of the girl who didn't take care of her teeth. I finished by producing the enormous set of plastic teeth and toothbrush I had in my bag and I demonstrated brushing techniques. The children laughed so much I wondered if they thought I was a visiting stand-up comic. I looked to Andresa for help and she quieted the class and told them about the toothbrushes and fluoride. I gave away toothbrushes, because no one here had money to buy them. We all went out to the school well, hauled up a bucket of water, and we brushed with fluoride as we shivered on the wet grass. Andresa didn't brush, but I did and the class laughed uncontrollably.

We returned to the classroom and I explained that I'd return once a week for twelve weeks with the fluoride. Andresa had the class thank me as I packed my bag. I went to the next class, Nidia's, and repeated my performance with the same result and then went on to Vilma's for my third attempt. That ended the morning session.

Vilma, Nidia and Andresa told me that most of the students didn't speak Spanish, but Guaraní, and that they understood little of what I said. They suggested that I do everything but the talking, and during the afternoon session each teacher explained in Guaraní what I had attempted to explain in Spanish. That worked much better. Many of the students left school that day with their first toothbrush.

15

Noemi's baby, Augusto, reclined on a cot, fat from layers of clothing, his eyes half shut and watery, and his face puffy and red. She thought he might have the measles, and, she, too, had a cold and a cough. I went to my room at the *Centro de Salud* and got her some aspirins and cough syrup and my copy of *Donde No Hay Doctor* (*Where there is No Doctor*) that Peace Corps had issued to each of us before we left training. We treated the rash with lotion I had. The baby had no fever and I surmised that he'd probably just had a reaction to a plant or animal. As it turned out, the rash and puffy face vanished within hours, but he did have a cold.

I left her house with an invitation for lunch the next day.

I went back to the *Centro de Salud* skipping around puddles and steaming cow dung to wait for Gaspar. He said he was going to fumigate some houses and I wanted to tag along to get census information on the houses.

Gaspar swaggered into the *Centro de Salud* and opened a storage closet and pulled out a plastic tank sprayer with shoulder straps and container of pesticide. He'd been trained by Dr N_____, he told me.

"Help me mix a tank, so we can get started, Marco."

"Sorry, Gaspar, but I'm not touching the pesticide, especially since I can't read the directions."

"Dr N_____ expects you to help fumigate the houses."

"I don't work for Dr N_____ and I told him I wanted nothing to do with it. I'm going along to get census information, not to help fumigate."

"He's not going to like that."

"I don't like pesticides."

"Well, that's okay today, until we can talk to Dr N_____ again."

"Gaspar, I have spoken to him. I'm not fumigating because I don't think it's safe. If you do, that's your business."

"Okay, let's go then," he said strapping the tank on his back and grabbing the container of pesticide. "I'll mix it at the house." He kicked the door shut and started walking. I shouldered my pack and followed him past the school, the sloping soccer field, and across open pasture to the house where I had asked directions on my first trip to Guazú Cuá. We walked without speaking, hearing only the sounds of killdeer screaming and cattle and horses snorting and blowing, and inhaling the pastoral scent of animal excrement, wood and charcoal smoke, and the occasional outhouse stench carried on the cool breeze. The windswept smoke from cooking fires from most houses.

At the house the family waited outside with their beds, cabinets, clothing, and some hand tools.

"*Buen día.*"

"*Buen día*, Señora. Is the house empty?" Gaspar said.

"Yes, it's ready."

"Okay, I just need some water to mix the formula," Gaspar said, looking around, expecting someone to provide the water.

"You can use the water from our well," Gaspar.

"Okay." He hauled his equipment to the well and drew several buckets of water, emptying each one in his tank. He then pried off the top of the container with the pesticide and poured some into the tank. He sniffed at the tank and added some more. He sniffed again and added more and sealed the container and the tank.

Meanwhile, I paced off the dimensions of the house and paced to the well, to the outhouse, and to the fence corners. I noted that the house had a straw roof, dirt floor, and rough, vertical wood planks for walls. A lean-to behind the house served as the cooking area. A covered breezeway split the house, dividing it into two rooms. I stepped back to sketch the house and to locate it on my map.

Gaspar donned a paper mask that covered his nose and mouth, secured the tank on his back, and said, "Watch how many bugs come out," as he entered one of the rooms and began spraying. And bugs did escape and mice and lizards, too. The stench of the pesticide was strong, so I told the family to move back, and they followed me to the road. At about this time Gaspar crossed the breezeway from one room to another to continue his fumigation.

Gaspar emerged from the house red-eyed and doffed his mask and dropped his tank on the ground.

"That's it," he said. "You can move your furniture back in."

"Maybe you should wait several hours. That's a strong poison, or at least it smells strong," I said.

"No, by the time they get the furniture in the smell will be gone," Gaspar said. "Go ahead."

The family began to move beds into the rooms. I protested, but the family wanted to be in its house and Gaspar said it was safe, and he was the one trained to fumigate.

I left and continued my census on my own. The doctor and Gaspar said the pesticide was safe and people believed them. I didn't know for sure, because I couldn't read the pesticide containers. I saw mice and lizards in their death throes after the fumigation. I had doubts about its safety.

I said my goodbyes and shook hands with everyone at the house and I walked along the road to the next house I hadn't visited yet.

I didn't get far before I met Jorge Encina on his way home from a morning of labour. He invited me to his house to drink *mate* and talk. I admired Jorge because of his differences. He lived in the community and the people respected him but his point of view varied from most, as did the views of Noemi's family. Jorge had lived in Argentina and had experienced another culture, not just that of Guazú Cuá. He hadn't lived his entire life under a dictator. And Noemi's father had never farmed, like most in Guazú Cuá, but had spent a lot of time travelling within Paraguay and to the borders of Brazil and Argentina before settling down. Few, if any, other people in the village had this sort of worldly experience.

I walked with Jorge who wore a shirt that had been repaired so many times that each sleeve differed in colour and had strong patches on the elbows. The collar, too, didn't fit the style of the shirt and the entire back was different than the front. The shirt had been reconstructed from several other old shirts. We reached his yard and opened the gate and walked to the house.

Outside, Julia squatted beside a pile of smouldering charcoal inside a small fire pit. She fanned the coals with woven straw fan until the coals burnt hot enough to boil water. She had a gourd filled with *yerba mate* and a silver *bombilla* ready on a table beneath the *galpón*.

Soon we sat in the mid-day chill with Julia serving the *mate*. She'd pour hot water in the gourd and pass it to Jorge and he would suck up the *mate* through the *bombilla* and would pass it back to Julia, who would refill the gourd and hand it to me, and, after I had sucked up my *mate*, she would take her turn. The gourd went around until we finished the kettle of water.

I told Jorge and Julia about the fumigation and that Gaspar seemed annoyed because I didn't want to participate.

"He thinks I work for Dr N_____, even though I've tried to explain my position. I want to work with him on his project, but he wants me, a volunteer, to complete his job so he doesn't have to pay anyone, or so it appears."

"He's already told people that there's no reason for you to be here because you're not doing your job," Jorge said.

"He told me he would report me to my supervisor. But I get along fine with the Germans. It's only him."

"People are talking about you spending a lot of time at Medina's, too. You're going to have to win the trust of the people. Many suspect you have something to do with the government or taxes. You are the first foreigner to live here. Germans have been here but only for hours, not to live. Let the people know what you are doing. People think you're doing the dental health program for Dr N_____, and if he hears about it, and he will, he'll probably take credit for it as part of his contract."

"But the teachers know. You know?"

"Sure, Marcos, it's not everyone. Dr N_____ makes a lot of promises and some people believe him. They think he's going to give them something. Those are the ones you have to watch out for. Others don't believe anything Dr N_____ says."

"I've got to get out of the *Centro de Salud* to break the connection with Dr N_____, but I do want to work with GTZ. It's a great agency and the Germans I've met are very dedicated."

"Some people think you're going to move to Medina's. This is a small community and you are the first North American to come here. Everyone knows who you are and everyone is watching you and talking about you. Some think you came here to escape the law or to get away from your family. Why would someone come all the way from North America to live here? That's what we've been hearing."

"I'm glad you told me."

"Go to the San Juan celebration at the school Saturday night. Get to know more people."

"You can watch Jorge walk on hot coals," Julia said.

I looked at Jorge.

"If you believe, you can't get burnt on the Day of San Juan," Jorge said. "You'll see."

"I will. I'm going now. Thanks for the *mate* and for the information," I said, standing and shaking hands with Jorge and Julia.

I went to the *Centro de Salud* to my room and sat at my table. I thought about what Jorge had told me and the situation I could be in if stories about me spread and the details of the tales changed with each telling, as I plotted locations on my map. Just then, Antonio's son, also named Jorge, burst in unannounced and said his mother had invited me to eat with them. I accepted and followed the barefoot boy through the chill, across soft road to the rich manure in front of Antonio's store, and then into the thick smoke swelling from Vilma's kitchen.

As director of the school, Vilma's relationship with me contributed to my success in Guazú Cuá, and I already sensed that she didn't approve of my visits to the Medina home.

"*Hola, Marquito, mba'e la porte?*" she said to test my Guaraní.

"*Muy bien, Señora. Ha u se,*" I said, meaning that I wanted to eat.

"You are learning," she said, and then, grinning and showing her silver and gold teeth, released a litany in Guaraní knowing I wouldn't comprehend a word, which brought mirth to the table from those who understood.

"Thank you for the invitation," I said, when the laughter subsided. "I appreciate it."

"Anytime. That's what we're here for," she said, smiling. She placed a platter of grilled beef on the table and a jar of salt. She then brought a platter of steaming *mandioca*. Antonio poured a mound of salt on the table in front of him and he snatched a hunk of beef. He rolled it in the salt and began gnawing on it. The rest of the family followed and I joined in the meal. Vilma brought a container of fresh-squeezed lemonade and a couple of glasses. She filled the glasses and we passed them around the table, two glasses for all. I tore at the meat and chewed until my jaw ached. I ate *mandioca* and left the meat. I drank some lemonade, wiped my mouth on a corner of the tablecloth, and excused myself, shaking hands and complimenting Vilma on the quality of the meal.

I spent the rest of the afternoon pacing off the last section of the boundary of Guazú Cuá for my map. When I returned to my room and plotted the last line on my map, I had a sixty metre bust in the boundary, meaning when I went all around the boundary of Guazú Cuá, I missed my starting point by sixty metres. I poured through my notes searching for my blunder and organizing information from the census. I had almost finished the map and census, and had only a few more houses to visit.

I had planned to go to Noemi's house for dinner, so I put my census away and started off just before dusk as the temperature began to drop. As I passed Antonio's house, and avoided stepping in a steaming cow pie, Antonio called me.

"Marcos. Marcos. Do you still want a horse?" I had mentioned that I'd like to have a horse, since it was the most common form of transportation in the area.

"Yeah, I'm still interested," I said, stopping in the field of manure in front of his house and store.

"Come on and try this one. I'll rent it to you."

"I'm just on my way to eat."

"It will just take a minute."

"Okay. I'll give it a try," I said, following him to the back of his house where he kept his horses and oxen.

Antonio wore his ratty straw hat with his pants rolled up to his knees as he padded along in his mud encrusted bare feet. He put the bit in his horse's mouth and pulled the reins over its head. He slung a blanket, a sheep's skin, and a saddle on his horse and cinched it tight. He then grabbed a bony grey horse by the mane and fitted it with reins and threw a blanket over its back. He held his hands together and motioned for me to put my foot in them so he could boost me onto the horse's back.

"Antonio, don't you have a saddle for this one?"

"I'll get you one. I have one, but not here. It's okay, it's a gentle horse." Antonio boosted me onto the horse's back and he climbed into the saddle on his horse.

"Let's go," he said, as he set off on his horse to the gate, which he opened without dismounting. My horse followed his and when we were out he closed the gate. He then set off at a trot across the common pasture land and my horse followed. I squeezed my legs against the sides of the bony beast to keep my balance and as we trotted I could feel the backbone of the animal sawing into me. Whatever Antonio's horse did, mine did. I yanked on the reins, I pulled to one side or another but the horse just followed Antonio's.

Darkness had fallen when we returned. I saw no light from the fire or candles at Noemi's. I dismounted and limped to Antonio's store. I bought some hard rolls and canned meat to eat in my room.

"What do you think? I'll rent it to you for 50,000 guaranies for a year. The saddle is extra."

"I'll think about it. I'd like to try it with a saddle." I left it at that and walked slowly back to my room.

That ride cost me a layer of skin and I suffered for it for days. I couldn't sit. I did finish my census and corrected my error in measurement during those days, though.

Meanwhile, people told me not to rent the old horse and if I really wanted one to get it from almost anyone except Antonio. People said that horse wouldn't last a year, anyway, and I'd pay more than he could get anywhere else. I hurt. My raw butt hurt and it hurt that I had trusted Antonio only to be taken advantage of.

The following days I limped about, over hills and through puddles and across a swollen creek to finish my census. I found the bust in my boundary measurement. I'd forgotten to record a distance along a fence line near the school. I hobbled back to the *Centro de Salud* to complete my census report. Few people understood the use of my census, though I tried to explain that it was to determine community needs. Many laughed and told me to just look around and I'd see the needs. There was no use for a census. For the most part, I could have done without the census, but in doing it I became familiar with the people and the area. The census itself didn't tell me much, but it was a formal study I could use to acquire grants.

I learnt that Guazú Cuá consisted of about fifteen square kilometres of land, half wooded and half either cultivated land or pasture. Four creeks cut through Guazú Cuá. Eighty-eight dwellings occupied the land with a population of 480, fifty-one percent men and forty-nine percent women. Fourteen percent of the mothers were single mothers. Twenty percent of the families owned small black and white televisions powered by car batteries. The batteries were hauled to Sapucaí, often on horseback, to charge. During the past year, two percent of the population moved to an urban area and at the time of the census, five percent planned to move to an urban area.

Of the eighty-eight houses, eighty-six percent were constructed of rough cut wood planks and forty percent were brick. Eighty-four percent of the roofs were straw, and the rest were of tile or metal sheeting. Seventy-nine percent of the houses had dirt floors and eighty-nine percent of the families cooked on open fires on the floor. There was no electricity or running water. Sapucaí, a town eighteen kilometres distant on a rutted dirt road, impassable during a rain, had the nearest telephone and postal service and transportation.

Sixty-three percent of the houses had wells with an average depth of five metres. Others used water from a neighbour's well or from a creek or spring. Fifty-two percent had outhouses without roofs; thirty percent had outhouses with roofs. Two percent had no outhouses at all, and the rest had sanitary outhouses. Sixty-five percent had a structure in which to bath, usually beside the outhouse. The rest bathed outside or in the house. Most houses had a garbage pit where the family burnt trash. There wasn't much trash, because people brought their own reusable bag to the market. Most food was sold in bulk without packaging. Bottles were reused for oil or honey storage. Dogs, pigs and chickens ate food scraps, so the burn pits usually had some paper, cans and broken bottles.

Before 1952, Guazú Cuá was uninhabited by humans, but there were many deer. Guazú Cuá does, after all, mean "the place of the deer" in Guaraní. Human habitation displaced the deer. The government sectioned Guazú Cuá into 200 by 500 metre lots, and in what would be the community centre, fifty by fifty and twenty-five by fifty metre lots.

Most of the people dedicated their lives to agriculture. They cultivated cotton and sugar cane to sell, and corn, peanuts, beans, and *mandioca* for personal use. There were plenty of grapefruit, orange, and lemon trees in the area.

Once a week, on Wednesday, the local butcher slaughtered a cow and sold the meat. People walked to the site and waited for hours to get a good cut. One family had a small lumber mill, capable of sawing logs into rough cut planks. Six families had small stores, of which Antonio's was the busiest and best stocked. Most sold only tobacco, beer and wine, sugar, salt, flour, and some canned goods.

The census told me what I already knew. Protected water sources and sanitary outhouses would be priorities. And since I lived in Guazú Cuá, I inevitably drank the water drawn from creeks downstream from where women washed clothes and cattle stood belly deep lapping at the water while urinating and defecating, and I relieved myself in roofless outhouses with floors of coconut trunks and three half-walls of coconut bark. A burlap bag or other material served as a door and a pile of corn cobs took the place of a squeezable, soft roll of paper. I tried boiling water, but it took too long to get a fire going and most of the water I drank was unboiled and shared with others at their houses, usually while drinking *tereré*.

The residents of Guazú Cuá would celebrate the Day of San Juan on this night, June 24. It rained but the faithful arrived at the school. A pit of coals sputtered in the rain. A believer cannot be burnt by fire on the Day of San Juan. A few men walked across the coals barefoot, and, apparently were not burnt. They didn't show pain, although a couple passed across the waning fire rather rapidly. A greased pole had been erected with something on top. I didn't stay to watch people attempt the ascent. As I sloshed back to my room at the *Centro de Salud*, I felt guilty about living in the best building in the community. It may have been the best, but it wasn't a home, and some of the undersized, one-room structures I visited that housed several people emanated hominess despite the material impoverishment.

I drifted into sleep thinking about moving, about my garden, about Antonio and the horse, and Noemi and her family, and how I might improve sanitation in Guazú Cuá.

16

I worked in my overplanted garden during the cool, cloudy morning transplanting seedlings from one *tablón* to another, but they lilted and looked at the earth rather than the sky. A few passersby complimented me on my model garden and said they'd be back for the harvest.

I left the garden and walked to the well, drew enough water to fill my five-gallon bucket, hauled it to the *Centro de Salud*, and used most of it to bathe. I changed into clean clothes, selected several bags of seeds, and started for Noemi's house. I had a variety of seeds from the Peace Corps office to give to families I visited during the census.

A Sunday soccer game had begun in the field across the street from the Medina's. *Don* Icho had a lumpy billiards table with ripped and patched felt and a few hand-carved cue sticks. It was the only game in town.

Bystanders wore jackets and sweaters as they shared bottles of beer and watched the soccer match on the sloping field. Others bet on billiards games. Noemi and her mother sold beer, *caña*, and food from a window near the billiards table. Men nodded to me, spoke to me in Guaraní, and shook my hand as I passed on my way to the house. I gave the bags of seeds to *Ña* Buenaventura as an offering and bought a beer. I sat with *Don* Icho beneath a *Paraiso*, a Chinaberry tree, and I told him about Antonio's horse. He laughed at the story.

"I want a horse because I want to ride to Potrero Arce to

visit Juan, the volunteer there," I said, "and to ride instead of walking everywhere."

"I'll ride with you Thursday, this week," he said. "I have a horse you can use and we have family in Potrero Arce. Come tomorrow in the *tardecita* (late afternoon or evening) to eat and we can talk."

"*Gracias, Señor*. I appreciate it."

Two girls approached and one smiled showing two silver teeth. She looked down instead of at me and said, "*Señor Marcos, por favor*, will you take a picture of us?"

"I don't have my camera with me, but why? Is it your birthday?"

"No. We just want our picture. We'll pay."

"I'll take your picture, but I'm not a photographer. I don't want to take everyone's picture just because they want a picture."

"Okay. We understand, but we'll pay."

It's only a photograph, Mark, I told myself. Why give them a hard time? I was the only one within miles with a camera. Some of the residents had never been photographed. I went to the *Centro de Salud* for my camera and returned to the Medina house. I shot a roll of film on young girls dressed in their best and I said I'd give them the photographs after I had them developed on my next trip to Asunción. I thought about Jorge Encina, who seemed so different. I thought that if he had a camera, he could take these pictures and profit from it. I had no desire to take the pictures or to profit from them. I thought I'd do some research on camera prices and developing and make a proposal to Jorge, because I trusted him. I could get him a camera, get his film developed, and he could pay me for the camera as he earned money. I wouldn't charge him anything extra since I went to Asunción every month anyway. What the prints cost is what he'd pay. I'd consider the idea.

At dusk the next day, wrapped for the cold, I returned to the Medina home and *Don* Icho invited me into the kitchen area to sit by the fire. The kitchen sat between one room of the house and the outhouse. The roof almost dripped creosote from years of a fire burning on the hard dirt floor beneath it. Wind passed through the vertical coconut bark walls that vines held erect. I ducked to enter the smoky area. Cooking utensils hung from wires suspended from the ceiling. A shelf held dishes on one wall and a table stood scattered with pots and pans. A fire smoked on the floor between a ring of rocks with a blackened grill of rebar on top. A sooty black kettle rested on the grill.

Noemi sat on the floor with her son, Pedro Augusto, on her lap. To her left sat Rodolfo on a short stool, sniffling from a cold. To his left sat *Ña* Buenaventura on the floor with Joel in her lap. She cooked and distributed the food in the pot. I sat beside her. *Don* Icho sat beside me wrapped in his ragged wool poncho. A young girl sat beside him. I thought she must have been his niece. She had black eyes and a broad white smile against her brown skin.

When I entered the room I felt as if I had been transmitted into the primal past. The family gathered around the popping fire with the *señora* fanning it, the bare feet, Noemi's dirty skirt and her bright face, *Don* Icho's toothless smiles, and *Ña* Buenaventura kicking red hot coals back into the fire. She scooped the polenta out of the pot and passed metal plates to everyone. I refused my plate even though I was hungry. I appeared at dinner time and didn't want it to seem as if I were looking for a handout. The steaming ground corn, onion, and cheese polenta smelt earthy. I had a bar of chocolate in my jacket pocket and I offered it around, but no one would take it. *Don* Icho said there were too many of them for the chocolate.

"Be ready early Thursday morning for the ride," *Don* Icho said.

"What time is early?"

"The first thing in the morning."

"But what time?"

"Dawn."

"Okay." I left feeling great as I ducked through the doorway, thanking everyone, my clothes stinking of smoke.

"Come back for morning coffee," I heard on my way out, "and bring the chocolate."

"I will," I said, as I thought about Noemi, always so happy, so in love with her baby, Pedro Augusto, yet a *campesina*, undereducated, but intelligent, so isolated and still full of life with nothing in common to my own being. Is it because of my loneliness here? If I had sat beside that primal fire with Noemi absent, would it have made a difference? And I wondered about her absolute acceptance of her lot in life, and I wondered if her family considered me a prospective son-in-law from the north? I enjoyed being with such people who smiled and joked and shared with no apparent ambition for enrichment beyond what their daily needs required.

17

"We can't travel in this weather," *Don* Icho said. "The roads are bad and we have to ford swift, flooded creeks. It can be dangerous. We'll let it go until the water drops and the roads dry."

"It will have to be after I return from Asunción, *Don* Icho," I said. "I have some days of workshops in Areguá at the Peace Corps training centre and then a conference in San Bernardino. I'll leave in a few days and should be back by July 9."

"Men make plans and God changes them," *Don* Icho said with a toothless grin. "We can go when you get back. I'll wait for you."

"I appreciate that, *Don* Icho. I'm looking forward to the ride." I waded back to the *Centro de Salud* to spend time writing letters home and reading issues of magazines I hadn't gotten to yet: *Newsweek* (provided by Peace Corps), *Harper's*, *The Atlantic Monthly*, and *The Utne Reader*.

***

I hiked out of Guazú Cuá Monday morning, July 3, a chilly, cloudy, dry day. I carried a big pack with just a few changes of clothes, so I'd have plenty of room to haul food on my return trip. I avoided mud and ruts and puddled water, yet mud covered most of my pant legs from the knees down. I had dry shoes in the pack. I passed few people on the road and most were going a short distance. Most of the locals would wait for better weather and a drier road, even if the wait lasted weeks. I didn't embrace the Paraguayan concept of time and I had to be on the move. I came with visions of assimilation although everyone within a thirty kilometre radius knew of me, one stalk of corn growing in the midst of a sea of wheat.

I waited for the bus on the shoulder of the slick, red dirt road and talked to a mother and her twenty-year-old daughter. They lived in Buenos Aires, Argentina, but were from Chircal. The young woman, Blanca, smoked and questioned me.

"Why are you here? Do you have a wife in the United States? Don't you think you'd like it more in Buenos Aires, or even Asunción, where things were more progressive? Would you stop at our house in Chircal on your way by to eat with us?" She flipped her cigarette butt into the road as the bus slid into view.

"I will stop at your house. Thank you for the invitation."

"We're here for a funeral. We'll be on our way back to Buenos Aires in about ten days."

The bus halted with screeching brakes and we forced our way up the stairwell and to the aisle. The bus lurched forward and jolted toward Paraguarí. When we arrived at the terminal and scrambled off the bus, Blanca grabbed my arm and asked me if I'd visit. I said I would and wondered about this local who had broken with tradition. "*Nos vemos,*" I said.

I wanted to get to Asunción quickly so I raced to Route 1, a few blocks away, to hail an express-type bus. One stopped and I ascended the steps and stood in the aisle with my backpack on until we reached the terminal in Asunción where vendors swarmed the bus offering gum, soft drinks, juice, beer, cigarettes, *empanadas* and *chipa* and trinkets.

I pushed my way off the bus, through the vendors and passed through the terminal. Outside I caught a local bus to the *Residencial Itapúa*. I paid for a room and showered and changed. I walked a few blocks to the Bavaria. I ate and had a few beers, but saw no one from my group.

I went to the *Residencial Itapúa* to sleep so I could get an early start to the hotel in San Bernardino for the conference. Peace Corps had made room arrangements and I roomed with John, who I had visited during training, and David, a member of my group.

I danced plenty to canned music during the conference. I met Sue, a volunteer living only eighteen kilometres from my site, but a journey of several hours because of the topography. We agreed to meet at her site on July 22.

During a talent show a few skits parodied the Peace Corps doctor, the nutritional value of Paraguayan food (so often fried), and volunteers who planned to divide their sites and then conquer.

After the conference I roomed with Steve at the *Residencial Itapúa*. The first night we went to a jazz bar with a gathering of people from our group. The next night I went out with Carl, an American I'd met at the University of New Mexico (UNM). He married a Paraguayan who had been working on a graduate degree at UNM. After his graduation, he came to Paraguay to live with his wife.

I enjoyed being with the members of my group. I needed to be with Americans and I needed to speak English and to talk about life in the *campo*. For some hours I didn't want to leave the comfort of my group and the English language.

I had to return to my site. Before I did, I signed out a Peace Corps issue mountain bike. I thought it might be better than buying a horse. My site coordinator said she could probably deliver the bicycle by Tuesday, July 11, depending upon the weather.

Before I left Asunción, I visited several embassies. Guazú Cuá was book poor and I wanted to start a library at the school. I asked at each embassy for books. Most gave me great books. The Italian embassy gave me books on Italian art. The American embassy gave me an anthology of American Literature, and a box of pamphlets describing life in America with photographs of people with equal opportunities, about Social Security, housing, and education. Most embassies gave me books on art or on the geography of whatever country it happened to be.

At this time the eighth person from our group left. She had been on the bus when it crashed. She said she hadn't been herself since. She seemed confused. She said she'd be back with the next group in February. Not one of the people in our group to leave was a person I expected to leave. There were now twenty-seven people in the group from the original thirty-five, and it had only been six months. That was twenty-three percent of the group gone during the first fifth of the experience.

***

A few days after returning to Guazú Cuá from the conference, I prepared for the day trip with *Don* Icho. Before daylight he had three horses saddled and ready. When I arrived we drank *cocido* and ate hard rolls. *Don* Icho checked the horses and we mounted for the ride to Potrero Arce. I mounted one horse and Noemi, Augusto, and her mother mounted another with *Ña* Buena at the reins. *Don* Icho and Joel rode the third horse. We had left the boundary of Guazú Cuá before the winter sun rose.

Don Icho led and I took up the rear as we followed an ox cart route through fields, across three streams, and through woods. We spotted some rheas along the way. Before we reached Potrero Arce I learnt that we were going to spend the night. I hadn't prepared and didn't even have my toothbrush. We rode into Potrero Arce after about a three-hour ride.

We stopped at the third house within the limits of Potrero Arce. The house belonged to Noemi's step-sister, Ana. She told us that Juan wasn't in Potrero Arce. He'd gone to Asunción. Ana prepared some eggs and polenta for us and we ate. The family spoke Guaraní and occasionally gestured toward me. When we finished eating, *Don* Icho stood and told me to follow him. We mounted the horses again and he led and I followed. We rode back toward Guazú Cuá and then turned onto a trail we had crossed. We followed it to a road and into Chauria, the neighbouring community.

Don Icho rode directly up to a house and stopped at an open window. I rode up beside him as he said, "*Buen día*," and I saw Kate, a member of my group, her mouth hung open, her breath held, and her features alarmed. She relaxed when she saw me, though her appearance changed from one of alarm to astonishment. As it happened, Kate rented a house from *Don* Icho's cousin, Margarita. He left me there for the day and the night, but left me the horse and told me to meet him between Chauria and Potrero Arce at 6:00 a.m. for the ride back to Guazú Cuá.

I spent the night at Margarita's house, but no local really believed that. Kate and I talked and drank a bit, but I don't think either of us was interested in a night together. I certainly didn't feel that way.

I awoke before dawn to a cold breeze blowing through the spaces between the vertical planks of the wall beside the bed. I dressed and made my way to the outhouse. I met my host hauling a bucket of water from the well. She dumped some in a basin on a small table beside the well.

"*Buen día, Marcos*. How did you sleep?"

"Fine, thank you."

"Use this water to wash up. There's a towel on that branch," she said, pointing to a towel hanging from a broken branch. "I'll have *mate* ready in a few minutes. My son will saddle your horse," she said, walking toward the three-walled kitchen.

"*Gracias, Señora*," I said, and I took a breath of cold, fresh air before I stepped into the outhouse. On my way to the kitchen I splashed the icy well water in the basin on my face and then dried with the thin towel. I sat in by the fire in the kitchen and drank *mate* with the *señora*. Her son poked in and said my horse was ready. I thanked Margarita and she walked with me to get my bag and then to the front of the house where the horse stood tied, blowing vapour from his nostrils and stomping with one foot. I untied the reins from the fence, pulled them over the horse's head, put a foot in a stirrup and pushed upward, but the horse turned and I stumbled to Margarita's laugh.

"I'll get you a chair," she said.

"It's okay, *Señora*, I can do it," I said as I struggled again. I sat astride the horse after a few attempts.

"Do you want me to send my son with you?"

"No. No, I'm fine. Thanks again for your hospitality, *Señora*, and please tell Kate that I didn't want to wake her. I'll see her in Asunción," I said, as the horse snorted and danced. I released the pressure on the reins and tapped my steed in the side with my foot. He sprang into a canter and I fell backwards on the saddle and pulled myself upright with the reins. He slowed with the pressure of the bit in his mouth yet he went where he wanted to go, which, fortunately, was where I wanted to go. I wondered if Margarita knew that I'd never saddled a horse as I bounced along the ox cart trail, my arms and legs wet with dew from brush the animal trotted through at times.

Dawn hadn't broken when we reached the road that ran from Guazú Cuá to Potrero Arce. I saw no one waiting. I looked across a field blanketed in very low fog. The horse turned right and trotted toward Guazú Cuá. I yanked on the reins and pulled to one side and we made a few circles before the horse realized who was boss and he walked toward Potrero Arce, stopping occasionally to sample leaves or pasture despite my kicks. We ambled up to Ana's house where the Medina's stayed. I slid off the horse and tied the reins to a fence and I went to the back of the house where I saw smoke.

I found the family drinking *mate*. They invited me in to share it and to have breakfast. I didn't mention that we were to meet an hour earlier on the road. Dawn broke to an overcast sky and a cool, gusty, breeze from the north. I sat with the family and listened to their laughter and their talk in Guaraní. I understood a word here and there and at times a few in a string, and Noemi translated when I appeared particularly puzzled.

Meanwhile, *Don* Icho saddled his horse and rode off. *Ña* Buena walked off with Joel. Noemi worked in the kitchen with Ana and I played with Augusto. I checked to see if my horse was still there every once in a while. Noemi wandered off with a machete and returned with an arm load of long leaves for the horses.

The morning didn't get any brighter. Darkening clouds built and the wind gusts from the north increased. *Ña* Buena returned with a bag full of cheese, bottles of honey, and *mandioca* flour. *Don* Icho trotted up on his horse.

"We have to go if we're going to beat the rain," he said, as he dismounted to prepare the other horse. *Ña* Buena and

Noemi gathered their belongings and hung bags from the saddles. *Don* Icho adjusted my saddle and tightened the cinch. "*Já ha,*" he said, leaping into his saddle and pulling Joel up after him. I climbed on the fence and stepped onto my horse. We said thanks and goodbye and rode toward Guazú Cuá with *Don* Icho in the lead. My horse took up the rear and did what the other horses did. It was 10:00 a.m..

We rode toward a black sky. We heard the clomping of the horse's hooves and distant rolling thunder. It rained. Rivulets grew to rapidly moving streams. Across a pasture, beside a stand of trees, stood a house and *Don* Icho left the trail and headed for the house. A couple with a baby sat beneath a *galpón* watching the rain. They invited us in and we rode the horses right under the roof of the *galpón*. It was an extension of the one-room bamboo and mud-walled house. A straw roof covered the room and an open area twice the size of the one room. I thought *Don* Icho knew the couple, but it was the first time they'd met. The woman of the house offered us *mate* and we sat talking and passing the gourd and listening to the water cascade off the straw roof into the red mud. We couldn't see the road from the house though the heavy rain. We stayed for two hours until the downpour subsided.

By that time I drank so much *mate* that I had to use a bathroom. Before we left, I asked to use the latrine and the gentleman pointed to the stand of trees behind the house. *Don* Icho accompanied me and we stood side by side urinating on a thick bed of wet leaves as water dripped on our heads.

We rode again and soon reached a creek like boiling milk chocolate running bank to bank. The horses waded across straining against the belly-deep current. We'd hardly pulled out of the creek before cold rain poured down. We rode to the next house and stayed until the rain let up. We wore soaking wet wool ponchos. They kept us warm as long as we didn't allow cold air in under them. We pushed on through the red mud and sporadic blinding torrents of rain, trusting the strength of the horses to ford streams as the current carried them downstream where we'd emerge from the water 100 feet below where we entered.

Arroyo Naranja-y separated us from Guazú Cuá. We had to cross just below the confluence of *Arroyo Pora* and *Arroyo Curuzú-Ruguá* where frothing, roiling water roared through its channel.

"We can't risk this one," *Don* Icho said, shaking his head. "I'm going for help." He spurred his horse and rode back the way we had come. We sat astride our horses beneath the canopy of a tree, shivering. We had to yell to one another above the roar of the torrent.

Don Icho returned with two men on horses. They'd brought ropes with them and they studied the creek from the bank. One eased his horse into the water and the current pushed it as it scrambled and swam for the opposite bank. Water washed over the horse's back drenching its rider. The horse struggled out of the water on the bank at a curve in the channel 100 feet below us. The rider guided the horse to the bank just opposite us and he threw a rope across to his partner. He made a loop to put around a rider's body. The other end was tied to the saddle horn of the horse on the opposite bank. The horse and rider could move to keep tension off the line.

"Let the horse do the work and keep the loop around your body and under your arms, like this," he demonstrated. "If the horse can't make it, or if you fall, we can pull you out."

I went first. I dropped the loop over my head and under my arms and nudged my horse forward. It didn't need prodding because it wanted to get home. I clamped my legs to the horse's back and clenched the reins and the mane. Only the horse's neck stayed above water as it swam and the current swept us to the curve where we hit the bank and the horse regained footing, leaping up to the high bank. I took the rope off and the *vaquero* threw it across the creek for the next rider.

Noemi crossed next with Augusto hanging to her neck and tied to her torso with a sling. She prodded the horse into the water and followed the same path as mine, emerging at the curve. We shuttled everyone across the creek the same way.

Don Icho spoke to the *vaquero* in Guaraní. They shook hands and we watched him cross back to his partner's side. We plodded into Guazú Cuá in darkness, drenched, drained, and chilled after the seven-hour ride. Rodolfo met us at the gate. We dismounted and he took care of the horses.

"Come back and eat with us after you get changed," *Don* Icho said.

At the *Centro de Salud*, I entered and lit a candle. I stripped off my sopping clothes, shivering as I rubbed myself dry with a towel. I dressed and returned to the Medina home.

The family sat around the fire drinking *mate*. *Ña* Buena fried some *mandioca* and eggs with cheese (*mandió xyryry*) and served everyone.

"I need a place to eat all my meals and I'd like to eat with you," I said to *Ña* Buena. That started a discussion in Guaraní of which I understood only a few words.

"We would like you to eat here with us," *Don* Icho said.

"Thank you. How much do you want for the meals?"

"Just bring the food and we'll cook it."

"I'd prefer to pay you for it."

"No. When you make your trips to town, bring back groceries."

"Okay. When will we start?"

"Now. Be here for breakfast."

I looked at the tired eyes around the fire. "I'd better go. *Gracias. Hasta mañana.*" I stood and ducked out of the kitchen and made my way to the *Centro de Salud*. I dropped onto my bed and slept until I heard knocking on the door.

"Marcos. Marcos. Go eat. The food is ready," Rodolfo said.

I roused myself, threw back my sleeping bag to the morning chill, and I fetched a bucket of cold water from the well to splash on my face. I walked to the Medina residence for my first meal, not as an invited guest, as I had been before, but as the result of an agreement to trade raw food for prepared meals.

The rain ended during the night and a chilly south wind blew billowing grey clouds northward. I stepped around puddles and mud as I walked to Noemi's for breakfast. Antonio and Vilma drank *mate* beneath their *galpón* near their store. School had let out for winter break.

"Come drink *mate* with us," Antonio said, holding up his gourd to show me.

"Thanks, Antonio, but I can't now."

"*Tranquilo*, Marcos," Vilma said. "There's always time for *mate*."

"Thanks, Vilma, but I have to go to breakfast and then to Escobar."

"You can have breakfast with us," Vilma said. "We're not going to charge you," she laughed.

"Thanks, but I'm having breakfast with the Medina family."

"You don't have time for your other friends now, Marquito? Didn't you just take a trip with them?"

"I rode to Potrero Arce with them, yes."

"Okay. Stop by and don't be such a stranger."

"Thanks, I will," I said as I made my way to the Medina home, just a few hundred feet from Antonio's.

I didn't mention the conversation with Vilma and Antonio as I sipped my *cocido* and sopped up egg yolk with a hard roll. I sat at the end of a table. My end had a cloth under my cup and dish. I had a napkin. The others at the table had their cups on the bare wood of the table. My dish and cup were different than the others. I had the only dish. The others ate rolls and only I had eggs. I'd been through this before.

"*Ña* Buena, why do I have a different serving than the rest?"

"You are the guest."

"Please, if I am to eat here, I prefer to eat what you eat in the same way you eat. I don't want to be treated special with a table setting that no one else has." That set off a chatter of Guaraní of which I understood little.

"My mother just wants to show her appreciation that you selected our house, out of all the others, for a place to eat," Noemi said.

"Forgive me, *Ña* Buena. I just don't want to be treated differently. I want to feel a part of the community, of the family, of the country."

Noemi translated my words into Guaraní for *Ña* Buena. She smiled and shook her head and disappeared into her kitchen.

"Thanks for the breakfast," I said. "I have to go to Escobar to call *Cuerpo de Paz* about a bicycle. It's a loan and someone is going to drop it off in Escobar or Paraguarí. What food can I get you while I'm there?"

"Meat and rice and vegetables we can always use," Noemi said. "Maybe some corn."

I walked along the muddy road with runoff still flowing from the storm. I made Escobar in three hours and went straight to ANTELCO to make a call. The woman who ran the outpost was out washing clothes, so I had to wait until she returned to the office.

When she arrived and opened the office, I asked to call the Peace Corps office in Asunción. The operator cranked her phone and connected with Paraguarí. She laughed and chatted a bit with the operator in Paraguarí and she, in turn, cranked her phone to connect with the operator in Itá, who, after a brief conversation, cranked her phone to connect with Capiata, and then finally with the Peace Corps office in Asunción. The operator handed me the phone and I spoke with Nanci, a third year volunteer. She said that she'd left my Peace Corps-issued mountain bike at Luci's office at the *Proyecto Paraguarí* in Paraguarí.

I caught a 4:00 bus to Paraguarí and made it to Luci's office by 4:30.

"Just stay here, Marcos. You have a room here and it will be dark soon. What's the hurry?"

"I know, Luci, but I have a lot going on in Guazú Cuá and I want to get back."

"So I've heard," she said. I let that go and said goodbye and pedalled to the market to purchase groceries. The air felt chillier and the daylight waned. I bought a couple of kilos of beef, some sausages, rice, pasta, bell peppers, and tomatoes. I stuffed it all into my day pack and rode out of the paved and cobblestoned streets of Paraguarí and onto the wet red road to Escobar, peddling as fast as my legs would move.

Halfway to Escobar, about six kilometres out, I had a flat tyre. The bike came equipped with a pump, but it didn't work. I walked the bike to Escobar in the dark and stopped at a store for something to drink. The owner pumped up my tyre and I waited to see if it would go down. It didn't. By then it was 8:00. I left and rode another five kilometres before the tyre went flat. I walked the bike home in the moon light and arrived after 11:00, filthy with mud and sweat and my disabled bicycle. I realized why most people used horses, oxen or their own feet for transportation.

Jorge fixed the flat on my bicycle. I presented my business prospect to him.

"I'm the only one around with a camera, Jorge, and I don't want to take pictures for everyone, even though people will pay for them," I said.

"Just tell them you don't want to take pictures."

"I have a better idea. I'll buy a camera for you. You can take pictures. When I go to Asunción once a month, I'll get your film developed. You can sell the pictures. You can pay me back for the camera when you make the money."

"Will you teach me how to use the camera?"

"Sure."

"How much will you charge me to take the film to Asunción?'

"Nothing. I'm going anyway. You just pay me for whatever it costs to develop the film. You can do all the parties and weddings and births. You'll make money every month."

"And you're willing to give me credit?"

"I will."

"I'll do it. Thank you for the opportunity."

"I have a selfish reason. I want people to stop asking me

to take pictures. I'll get the camera and film the next time I'm in Asunción. Now I have to take pictures at the party."

I walked to *Ña* Buena's fiftieth birthday party with him and Julia. The Medina's slaughtered a cow and cooked it over a rectangular open pit of coals. The meat hung from stakes and fat dripped onto the coals. Glasses of sweet red wine mixed with soda passed from hand to hand, from mouth to mouth. I didn't like the wine or the mix, but I imbibed and I partook in the breaking of bread. I had the only camera in the area, so I shot picture after picture, loaded on wine. I tried to get next to Noemi until I felt too spirited, so I excused myself and sauntered back to the *Centro de Salud*.

18

The night sky shined with stars and I sat alone gazing at constellations. *Don* Caceres, Julia's father, saw me sitting alone as he passed on his way home from the party.

"I'm going to Potrero Jara tomorrow on horseback. If you want to follow me on your bicycle, I can show you where you can leave it when you go to Asunción."

"I'd appreciate that."

"I'll be by at sunrise."

The next morning I rode with *Don* Caceres to Potrero Jara. His great grey horse clopped along at the speed I could walk, so I had trouble going as slowly as him on my mountain bike. I rode by his side and he introduced me to everyone we passed on the road. By noon we reached the main road and we stopped at a house. *Don* Caceres would leave his horse there. He introduced me to the owner of the property and I got permission to leave my bicycle in the house while I was gone.

Don Caceres, myself and several others, stood in the shade by the roadside waiting for the bus. The first blue bus blew by us at 1:30, full. People hung from the doors, the bumper and on the roof. At 3:30 the next bus passed—full. I decided to walk. *Don* Caceres couldn't accompany me because of his heart. He may have had *mal de chagas* and was on his way to the clinic in Asunción for tests.

I hiked about three kilometres to Escobar. A car with a man and a woman stopped for me. The woman spoke

English. They drove me to Asunción and I got a room at the Stella.

I went to the bank for my living stipend, had my film developed, and bought meat, grains, and vegetables for Noemi's family. I picked up my mail at the Peace Corps office and then ate and had a few beers with some other volunteers at the Baviera.

The next morning, I took a bus to the terminal and caught another bus to Potrero Jara where I had left my bicycle. I rode the last fifteen kilometres to Guazú Cuá, arriving at 1:00 p.m. I rode straight to the Medina's house and dropped off the food and I gave Noemi the pictures from the party. I rode to the *Centro de Salud* and went to the well to draw water to bathe. Jorge saw me and he came by for his pictures. He said he'd bring a load of fertilizer from the hills and spread it in my garden. Maybe my garden has a chance, I thought.

I decided to fix up the old house and move out of the *Centro de Salud*. I walked over the hill to find Francisco, the *mixto* driver, and I ordered and paid for 500 bricks. I felt committed to move after ordering the bricks. On my way home I stopped at Osorio's house and found him between his house and field with an ox cart loaded with corn. I helped him push the cart up to his house and we unloaded it. We sat beneath a tree afterward and drank *tereré* and talked. I told him about my plans to move and told him about my suffering garden.

"There are better places to live. You could build a place with the money you'll spend fixing that place up, Marcos."

"You think so? It needs a lot, but the location is great."

"Think about it before you commit yourself. You're going to be here a long time, right?"

"Two years."

"If you put money into the place, when you leave, the owner profits. If you build a place, you can sell it."

"You're right. I hadn't thought that far ahead. I'll think about it. I'll take another look at the place on my way home."

"Let me give you a bag of chicken shit for your garden. If that doesn't help it, move your garden on the hill. That's where you find good soil."

I left Osorio's with a heavy sack of chicken manure on my shoulder, thinking about what he said. I thought I'd be helping the people, but the people were helping me. They knew how to live there. I didn't.

I stopped at my prospective home and dropped my sack of fertilizer. I hadn't realized how much work it would need. It wasn't much more than a stall with a porch. I could fit my thumb in the spaces between the vertical wall planks. One frame for a window looked out over pasture, but the shutter had been taken. No fence stood to keep out farm animals. It would be a project. I shouldered my sack and walked to the *Centro de Salud*. I saw that Jorge had already spread fertilizer so I added Osorio's as well.

I took a bucket bath, changed and went to Noemi's for supper. I sat talking to Noemi, *Don* Icho and Rodolfo while *Ña* Buena cooked. When I went to Asunción I learnt that limited funds were available to pay a tutor for Guaraní. I offered the job to Noemi and she accepted. I then talked about my brick order and plans for the house.

Don Icho began speaking in Guaraní and the rest of the family joined in, each looking at me as they talked. I concentrated and tried to pick up a few words, knowing I was the subject of the conversation. They went back and forth for a good twenty minutes.

"Why don't you live here, Marcos?" *Don* Icho said. "We

have a room we could give you. You have your meals here. *Ña* Buena and I want to move to Paraguarí soon. It will save you the trouble of fixing a house that someone else will profit from."

The Medina house consisted of three buildings. One closest to the road, just across the street from the soccer field, was newest, two rooms, and built of brick with a tile roof and a tile floor. Beside it stood the main house of wood with a brick floor and a straw roof. Sheets of tin formed a gutter between the two buildings to catch rain water and a length of coconut bark channelled it to the well. Doors were aligned to pass from one building to the other beneath the rain gutter. Next to the main building stood the original house constructed of rough-cut vertical plank walls and a straw roof and a dirt floor. The roof extended past the one room at either end forming an outdoor sitting and dining area at one end and a partially closed cooking area at the other end. Behind the cooking area on the fence line were two stalls of brick with a tile roof. One was the latrine and the other the bathing stall, or the *bañadera*. *Don* Icho offered one room in the brick building.

"*Gracias*," I said. "I'll think about it for a few days." I left pleased yet worried. I remembered Jorge mentioning something to me about spending a lot of time at the Medina house and Vilma made a comment or two.

Men are men and women are women and tradition suggested that the two should not meet without a chaperone unless and until marriage. Notwithstanding, every other house had a *madre soltera*.

The custom here was that the male, the *novio*, visited his interest, the *novia*, at her house when the parents were present. Some families held to certain nights a week for visits. When a dance occurred, family members accompanied the couple. A man and a woman would never be left alone because, apparently, they would take the opportunity to copulate, and the chances of a condom being in the area would be slim. I heard some men say that they wouldn't wear a hat on their penis anyway. Single mothers abounded, and often married men planted the seeds. Some men had a primary family and maybe one or two others.

I left wondering about the offer to stay at the Medina residence, and asked myself again why the family was different and why some of the neighbours did not like their ways, though I had felt comfortable with them. I reflected on why I had spent so many nights gazing at the southern sky alone while flossing my teeth.

A spacious distance has separated me from the locals, I observed, but can that expanse ever be bridged? Will I spend all my time here as the *Norte-Americano-Aleman* as many call me, unaware of geography, or the *rubio*? Is it possible for a Peace Corps volunteer to meld into a community? Or will the foreigner remain a drop of oil in water?

19

In August, Areguá-1 gathered for a week of in-service. I left Guazú Cuá with some personnel from GTZ who had come to celebrate the first anniversary of the *Centro de Salud*. They provided meat and Antonio and a few others built a fire pit and hung slabs of the meat from stakes over the fire. Men played cards. Vilma sold beer, soda, and *caña* from an ox cart. The GTZ staff vaccinated children and saw dozens of adults with health concerns. I hadn't yet seen as many people at the *Centro de Salud*. In the afternoon, after all patients had been treated and vaccinated, I squeezed into the back of a white four-wheel drive Mercedes Benz with a driver and three nurses from GTZ and rode to Asunción. From there I caught a bus to Areguá.

The members of Areguá-1 stayed at the same homes where we lived during our primary training. I hadn't seen many of the volunteers since training, and the group was noticeably smaller. One died and seven others went home. I heard that most volunteers who leave the Peace Corps do so during their fifth or sixth month at their sites. I'd been at my site for three months now.

Most members of Areguá-1 lost weight and many wore shoes stained red from the Paraguayan soil, but we still looked unmistakably American. The training contractor, CHP, separated us into small groups according to our needs. I studied Guaraní and Spanish all week. Torrential rains kept the air fresh as it drummed on roofs and gurgled along streets on its way to feed *Lago Ypacaraí*.

I had time during the week to reserve a projector and a generator at the Peace Corps office. I rented a few reels of film from a business in Asunción. The films would be the only entertainment in my site and would generate funds for projects. I bought an *horno* and a *plancha* to use for my first *fogón*, a brick wood-fired stove with an oven. The *plancha* formed the cooking surface and the *horno* the oven. A *fogón* allowed women to cook standing up rather than bending over a fire on the ground in a smoke-filled room. It consumed less wood and the smoke escaped through a chimney.

Blanca, the Peace Corps coordinator for my area, agreed to take the projector and generator and my *fogón* parts to *Proyecto Paraguarí* and I'd find a way to haul them to the site from there.

I visited SENASA, the national environmental agency with which I worked, and arranged to get a hand pump for the school well. An American agency would donate the pump, but to acquire it I had to have it delivered to SENASA and have a letter signed by Vilma, the school director. I still needed the letter and then a wait for the pump. Packing it out of Asunción and to the site would be a challenge.

After the in-service I caught a packed bus to Paraguarí and bought food at the market. My pack weighed about sixty pounds. I went to the *Proyecto Paraguarí* and found Rosa. She gave me a ride to Potrero Jara, but she wouldn't leave the main road to drive toward Guazú Cuá because of the mud and, even without mud, she would have had to drive out alone. At 3:00 I lifted my pack out of the bed of the truck and strapped it to my back. Rosa turned back toward Paraguarí and waved.

I hiked the fifteen kilometres to Guazú Cuá in the clean, moist air, past dense wet woods, pasture, cultivated fields, grazing cattle, stands of bamboo bent with the breeze, and straw-roofed houses. I plodded through squishy red mud and huge, deep puddles and greeted barefooted women with baskets or buckets balanced on their heads, and men wearing straw hats with long knives in their belts driving oxen or sharpening machetes.

The road climbed or dropped but never levelled. The pack and the mud slowed my pace and I reached Noemi's house tired, wet and mud-splattered. I emptied most of the contents of my pack on a table—meat, flour, beans, peas, cheese, coffee, jelly, *chipa*, and some tomatoes.

"*Gracias, Marcos,*" *Ña* Buena said. "Sit down and eat some *bori-bori*."

I sat and Noemi brought me a bowl of soup of vegetables and meat with dumpling-like small oval balls of corn and cheese and boiled *mandioca* on the side. I ate and talked about the movies and the oven. I'd have to wait until it dried before I could get the equipment to the site. A thin young girl wearing layers of multicoloured clothes appeared, sat, and watched me eat. Her brown hair fell around her shoulders and she stared through intense brown eyes.

"That's Fatima, my cousin," Noemi said. "She's my mother's brother's daughter. She came from Durazno and she's going to stay with us to go to school. (I'd seen her at the cooking fire one night). There's no school in Durazno yet so she's already behind. She's nine now."

I spoke to Fatima but she just sat in her layers of hand-me-downs and stared with her big eyes and smiled. I finished the bowl of *bori-bori*, thanked *Ña* Buena, and went to the *Centro de Salud* to wash and rest.

A few days later, Rosa and a helper, Francisco, brought my generator, an extension cord, projector, reels of film, and my parts for a *fogón*. Antonio walked over to see what

Rosa brought. He helped us unload the equipment. I invited Rosa and Francisco to stay and drink *tereré* and to have lunch, but she said she had other stops to make and she didn't want to get caught in the rain on the road out. I looked at the puffy white clouds floating northward across the blue sky in the late morning sun and saw no sign of rain.

"It's not just rain, Marcos," she said, glancing at the sky, "but the road is a mess, *ivaí tape*. Deep ox cart wheel tracks can swallow this truck, and we came halfway here sideways there's so much mud." She gestured toward the *Proyecto Paraguarí* truck sheathed in drying globs of mud.

"Okay, Rosa. I appreciate your time. Be careful on your way out."

"Good luck with your projects," she said, getting into the driver's seat. Francisco climbed into the passenger's seat. Rosa started the truck, backed out, beeped, waved and drove off.

"I wonder why he lets her drive," Antonio said, watching the truck splash through a puddle. "I wouldn't, would you?"

"Of course. She's probably a better driver, and Francisco is a labourer. Rosa is an engineer," I said. Antonio shook his head.

We dragged the equipment into a room. I told Antonio I'd stop by to talk to Vilma about having a movie night. He left and I took the parts for the *fogón* and carried them to the Medina house. In doing so I had to pass Antonio and Vilma's house. I decided to build my first *fogón* at the Medina's, since I ate there and because many people stopped there to play billiards or to drink beer, especially during a soccer match, since the field was just across the street. It would be an advertisement, and I could use it to bake bread as respite from the rock hard rolls I usually gnawed.

I told *Don* Icho that I'd build a *fogón* with the bricks I had ordered from the *mixto* driver, when he got around to delivering them. I explained how it would work and showed him where we could place it in the cooking area. *Ña* Buena muttered in Guaraní and, although I didn't understand the words, I felt her scepticism.

I spotted Fatima holding Augusto. She stared at me from outside. I waved and she vanished. I said goodbye, left the materials there and went to talk to Vilma.

"Why don't you build a *fogón* for the school instead of for your *novia*?" Vilma snapped as I approached.

"Vilma, I don't have a *novia* and I will build a *fogón* for the school, but I want to build a kitchen for it first."

"Then you're the only one that knows you don't have a *novia*! I thought you were here for the community," she spat. I hadn't heard her use this tone.

"I came here to talk to you about an activity to raise money for the school, Vilma. I have everything I need to show movies at the *Centro de Salud*. Will you help me?"

"For the school? You want to build a kitchen?"

"Yes, and I have a pump for the well, too. It will take me some time to get it here, though."

"Okay. What about Saturday night? Antonio can sell food and drinks and we'll charge 500 guaranies per person for the movie."

"How can we let everyone know?"

"Go to the radio station in Paraguarí and have them announce it. I'll write it for you. They'll charge you for each time they mention it, but it's not much."

"I'll go on my bicycle in the morning."

"You can get benches from the school."

"I'll get them. I have a letter I need you to sign, too. It's for the school pump."

"Let me read it first."

"I will, but it's in English. It's an American agency that will donate the pump. I'll stop by later with the letter."

"Bring me the names of the movies, too, so I can write the advertisement." Her tone of voice softened.

In the morning I showed Vilma the letter and she signed it. She gave me the notice for the radio station. I wheeled out of Guazú Cuá on my mountain bike and took a road that went directly to Paraguarí, thirty kilometres. It should have been a road, but only the right-of-way existed bounded by fencing. Groups of ox carts used the right-of-way to haul cotton, sugar cane, charcoal, *mandioca*, tobacco, or other goods to the market in Paraguarí. Once they sold their products, they'd load the carts with merchandise for the *almacens*. A horse trailed behind each cart so the driver had transportation in town.

Between Guazú Cuá and Paraguarí, along the right-of-way, cattle grazed in pastures, rheas stood in fields, and monkeys and parrots screamed and screeched in the woods. A few streams cut across the right-of-way and one major creek. At times a dozen or so ox carts and drivers camped beside the creek, with their oxen and horses hobbled nearby. I made the round-trip to Paraguarí without passing one person along the right-of-way.

I rode straight to the radio station, locked my bike outside, and went into the one-room station. The announcer took my message and I paid to have it read twenty times over three days. He read it once while I was there. People used the station to send messages to relatives, or invitations to events in areas lacking telephones. Someone in the community would hear a message and it would spread like lice so all would soon know. In this manner people in Guazú Cuá and nearby communities heard about movie night.

I ate at a restaurant in Paraguarí and then I purchased a container and gas for the generator. I strapped it to the rack on the back of my bicycle and rode back to Guazú Cuá along the right-of-way, a sixty kilometre round-trip. I found one major defect with the bicycle. Mud stuck to the tyres until the tyres grew so fat they wouldn't roll. I'd stop and scrape the mud off both tyres and from the frame so I could continue. As long as I avoided mud, the bike rode fine, but once I hit mud I'd soon have to stop completely. I'd never seen anyone scraping mud off their horse's legs. The mountain bike hadn't been designed for Paraguayan mud.

As I rode, scraped mud from the tyres and listened to monkeys mocking me, I thought about what Vilma had said. When I left Guazú Cuá for Asunción, I left my radio with Noemi. I loaned her books. I ate at her house. I played with Augusto, her fourteen-month-old son. I'd travelled out of town with the Medina family. And I have had the woman on my mind. Why shouldn't Vilma believe I have a *novia*? Maybe she's right. By local custom it may be apparent to everyone, while I saunter through time oblivious to what others see. It was like realizing my zipper was down the entire time I danced at a party. I rode and I scrapped and I pushed and I rolled into the *Centro de Salud* a muddy mess. I'd have to take my bike to the creek to wash it.

I sat in the breezeway of the *Centro de Salud* and drank *tereré*. I planned my movie set up. I'd set the projector near the door on my side of the building and project onto the white-washed wall of the clinic-side. I'd arrange the benches in rows. I'd ask Vilma to collect the entrance at the gate. I didn't want to handle money. Antonio could set up beside the projector to sell his drinks. I'd run the extension cord from the projector through my side of the building and out the other side where I'd put the generator. The building would buffer most of the noise.

Saturday came and I set up and practiced with the reels and the projector. I had a *Three Stooges* film, a circus film, and one of *Popeye* cartoons. I threaded the *Three Stooges* film and had it ready. At dusk families began to arrive. Young men tied their horses to the fence. Some people came several kilometres to movie night. There weren't enough seats, but no one seemed to mind. Women and children sat on the benches and the men stood in the back drinking beer and smoking hand-rolled cigars. A woman showed up with a basket of *chipa* to sell. I told her we were raising money for the school and she couldn't sell her *chipa*, but she smiled toothlessly and acted as if she had no idea of what I said as people crowded around thrusting money at her for *chipa*.

The audience clamoured for the movie, so I ran around the building, started the generator, ran back to the projector and started the reels rolling.

At midnight, I put the projector in my room and got someone to help me lift the generator onto the landing of the *Centro de Salud*. Most of the crowd had gone, asking as they left when I would show movies again. A group of beer drinkers remained with Antonio. A few stood laughing and urinating on the fence.

"What do you want to do with the money, Marquito?" Vilma said, holding a cloth pouch in her hand.

"Why don't you hold it? It's for the school and it will be safer at your house."

"Okay. I'll be the treasurer then," she said. "Have a good sleep." She stuffed the pouch in her jacket pocket and walked out of the *Centro de Salud* and to her house. I said good night to Antonio and refused offers of beer. I could hear Antonio trying to get rid of the drinkers as I fell asleep.

20

On Wednesday, August 30, 1989, the feast day of *Santa Rosa de Lima*, and the Day of the Police, many members of the community gathered at the *Alcaldia*. Residents gave gifts to the *Alcalde*. Smoke rose from a cooking pit and slabs of beef and mutton hung from stakes and dripping fat hissed on the coals. Men passed glasses of beer and *caña* from one to another as they talked and smoked hand-rolled cigars and cigarettes. Three men, *Don* Juan, Desi and Silvio, sat on a bench in the midst of the revellers picking and strumming their guitars while singing songs in Guaraní. Women placed baskets of *chipa* and *sopa paraguaya* and bowls of *mandioca* on tables. Men turned the meat on the stakes and raked the coals of the fire.

I watched *Don* Juan plucking his nylon guitar strings, his head tilted up, a broad-brimmed blue cloth hat on his head and his full black moustache moving with his mouth as he sang.

"Marcos! We have to talk," Vilma said, seizing my shoulder and spinning me toward her.

"What is it, Vilma," I said, meeting the ire in her eyes. The cloth money pouch hit my shoulder and dropped to the ground.

"So you think I'm dishonest? You take the money. Let that gossipy bitch *Ña* Buena—what a name—hold it. She's telling everyone that I'm stealing the money from the movies. I won't help you again and neither will Antonio. Find another *cantina*. No one will help you. See if your *novia* will help! And I'm going to report you to Peace Corps. The family you eat with is not part of this community. Do you see them here? We don't need you here either," she spat, turning and storming off. The music stopped.

I heard muffled laughter and turned to meet a dozen pairs of grinning eyes. *Don* Juan, Desi and Silvio started another song. I picked up the money pouch and started walking to the *Centro de Salud* to store it there.

"Marcos! Wait for me."

I looked and saw Jorge Encina walking toward me with a floppy straw hat on his head, his best, patched clothing, and a grin glowing from the black stubble on his dark face. I waited and then we walked together.

"Don't let her bother you," he laughed. "Some people won't attend school functions because they don't trust the teachers. Some won't attend the police benefits because they don't trust the police. Some don't go to the *Centro de Salud* because they don't trust Dr N_____ or the technicians."

"I don't need problems with the school director, Jorge. I need to work through the school."

"That's the way she is, Marcos. She'll get over it. Plenty of people will help you. I'll help you with the movies and we can set up our own *cantina*. I'll carry it in on my back if I have to."

"Thanks, Jorge. I wonder why she's going to complain about me. I haven't done anything to her." We reached the *Centro de Salud*.

"Do what you're going to do. I'll wait for you. We can go back to eat."

"Where's your camera?"

"Julia has it. I'll take pictures later, after everyone has had some drinks."

"Are you still going to work with me on the *fogones*, even though the first will be at the Medina's?"

"Marcos, I don't bow to anyone here. If you have work for me, I'll do it. I already know you, because you got me started with the camera. I know you want to help. But you came to a place where the people don't want to change. That's why Stroessner ran the country for so long. He led the people like marionettes and if one didn't dance the right way, he cut the string."

"Okay, Jorge. Let's build the first *fogón*. It will be a model. Several others want *fogones*, too. Nidia, for one. She can be the second."

"*Mañana?*" Jorge said.

"Can we?"

"I'll be there in the morning."

I locked the money pouch in my room and went back to the party with Jorge. No one said anything about Vilma, but I felt eyes boring into my head. Antonio forced a smile and handed me a bottle of beer. Someone gestured to a seat at a table. Meat, *mandioca*, *chipa*, *sopa paraguaya*, piles of salt, and bottles of beer covered the table. We dipped the meat in salt and ate with our hands, throwing bones and scraps over our shoulders to the dogs. We used the edge of the tablecloth to wipe our hands and mouths. I yanked at the sturdy meat with my teeth, some dripping red and some scorched. I filled up on the *sopa paraguaya* and excused myself so another could take my seat.

I saw women eating and sitting in the shade of a lapacho tree heavy with brilliant lavender blossoms. Vilma spoke gesturing toward me and several heads turned. I recognized no face except Vilma's.

Gossip said that during school Vilma would have her students, and those of the other teachers, go to the woods when they should have been in class to gather armloads of dry branches, which Vilma used to fuel her beehive oven to bake cake that she sold to the same students. That money never found its way into the school fund, so said the story. Vilma had been suspected of pilfering school funds and a group wanted to oust her, including Noemi's family.

I wrote to my sister, Fran, about that day: " … she went on slandering Noemi and her family. She told me I'd better find another woman and had the audacity to suggest one. She said she was going to notify Peace Corps to tell the agency I'd gone astray. She did, in fact, send a telegram."

I shook hands with all the men at the tables and listened to one more song from *Don* Juan, Desi, and Silvio. I gave a thumbs up to Jorge and I walked back to the *Centro de Salud*. I packed the reels of film and a change of clothes so I could walk out the next day to get my living stipend, my mail, new reels of film, and develop film. I needed time away to think and to ask advice from other volunteers who may have had similar problems in their sites.

After the party, Jorge brought me a couple of rolls of film. He had a list of some shots he wanted doubles or in larger sizes.

"Are you sure this is okay, Marcos? I know it's just extra work for you."

"No, Jorge. I'm glad to do it. I have to get my own film developed anyway. It's not extra work. And if you're gaining a bit from it, it's worth the time to me. We can build the *fogón* when I get back."

"I'm not going to forget you, Marquitos. You'll see. I'm not anyone, but everyone wants me to work because I'm good. That's how I'll help you. Don't worry about Vilma. And Antonio is okay, he just doesn't wear the pants in his house. *Entiendes?*"

"So I gathered."

***

Two days later, Friday, September 1, 1989, I arose early and slipped out of Guazú Cuá before anyone knew the difference. Or so I thought. Every woman milking a cow saw me leave.

I hiked to Potrero Jara and waited for the bus. The bus passed every two hours and sometimes it was too full to pick up more passengers. Luck followed me that day. The porter threw my pack into the baggage section and I boarded the bus and paid to the terminal in Asunción. People dozed in every seat and men filled the aisle. I squeezed my way to the middle of the bus and reached for a handhold as it lurched forward. I stood for three hours until we arrived at the terminal.

At each stop, vendors swarmed the bus outside the windows selling drinks and *chipa*. Occasionally a *chipera*, a woman who boarded busses to sell *chipa*, would work her way to the back of the bus yelling, "*Chipa, rica chipa. Chipa, rica chipa!*" These vendors always sold and would get off at the next stop and board another bus. All vendors wore low-cut blouses and short skirts. I always let them pass, but they made an effort to rub against me on the crowded bus. Some said they sold more than *chipa*.

In the terminal of Asunción, after retrieving my pack and pushing past the vendors who accosted me with: "Watches? Belts? Medicine? Women?" I caught a local bus into town.

I called Carl, my acquaintance from the University of New Mexico. He invited me to stay at his apartment he shared with his wife, Luci. She had left on field trip so he had the apartment to himself.

I ran errands in Asunción. I gathered my mail, got my living stipend, bought rubber boots for myself and for

Noemi, changed movies, developed film, and bought some food. I rode busses and walked along busy streets lined with fruit, palm, and lapacho trees, some blossoms a vibrant yellow and others radiating purple.

I went to Carl's apartment in the afternoon, dropped off my goods, showered, and we went out to eat and drink. Carl said he'd like to get out of town. Luci would be gone for a few days. She was on a wildlife count or something of that sort. Carl and Luci both earned graduate degrees in biology, but Luci held Paraguayan citizenship and already had a management position with a government agency.

After dinner and a few drinks, I invited Carl to spend some time at my site. He had a Volkswagen Beetle. I told him we could drive to the *estancia* and leave his car there under care and walk into Guazú Cuá. He agreed, yet doubted me.

Notwithstanding, in the morning, we loaded my gear into his VW as well as his pack and we drove to the *estancia*. A guard stopped us at the gate and after a short interrogation, told us where we could park. We drove there and there was nothing around, just a fence.

"Don't worry, Carl, your car is okay here."

"How do you know?"

"I know because the people here know the families I'm involved with in Guazú Cuá."

"Right. I'm holding you responsible."

"Don't worry about it, Carl. These are good people."

"There are no good people, Mark. But if you are willing to take the risk, Okay. It's your ass."

"Then get your ass out of the car. We still have to walk."

We slipped through the barbed-wire fence on the boundary of Guazú Cuá and the *estancia*. Carl carried a small pack and I carried a pack with three reels of film, as well as my clothes and pictures and food.

"This is quite pastoral," Carl said, as we passed fields of corn, *mandioca*, tobacco, and cotton. Cattle grazed in the common pasture. Monkeys screeched from trees as we walked. Barefooted children peeked at us from behind trees. The purples and yellows of the lapachos splashed against the backdrop of greens. Women lowered their eyes as they walked by us on the trail. Men called, "*Adios*," as we passed.

"It's more pastoral than you know, I think, Carl."

"Do these people know you?"

"They know me or they know of me or they know what they have heard of me. Yes. They probably know more about me than I do."

"And so they will know me?"

"If not more so, since you'll only be here for a couple of days."

We arrived at the *Centro de Salud* and Carl and I toured my temporary home. I gave him the room next to mine. The rooms were meant to be hospital rooms, but it hadn't worked out yet.

"There isn't a bed here."

"We can borrow a cot. Wait until tonight, Carl. I have some great films here. Let's go get a drink and something to eat. We'll ask about a cot."

We walked past Antonio's and waved and continued to Noemi's. The family poured out to meet Carl, or Carlos.

"Carlos is a friend from New Mexico in the United States. He lives in Asunción and he's going to help me with the movies tonight."

Everyone shook his hand, including Fatima.

"How old is he?" *Ña* Buena said.

"I'm thirty," said Carlos.

"Thirty?"

"Thirty?"

"*No puede ser!*"

"Really," said Carlos.

"We thought you were twenty," Noemi said.

"Sit down, sit down," *Ña* Buena said, pushing a chair into Carl's legs. He sat.

"*Ña* Buena, *por favor*, give us a beer and a glass of *caña*." She served the beer and rum and watched us. "Carlos, come see where we're going to build a *fogón*."

We walked to the cooking area and *Ña* Buena sat cooking on the floor beside where the *fogón* would soon stand. "Well, there's where it will be." I explained to Carlos how it would work if it were used. We went back to the table.

Carl and I spoke English and we shared a few beers and several shots of *caña*. I told him everything I knew about the community and about the rift with *Ña* Vilma, or Vilma as I called her.

"I didn't know this would be so much fun," Carlos said, swirling *caña* in his glass.

Ña Buena served us a meal of beef with a fried egg and onions on top and with *mandioca* on the side. We ate and Carlos paid for the meal and the drinks. We went to the *Centro de Salud* to set up the projector and the generator for the night's entertainment.

People came from kilometres around. They wanted to see the movies or they wanted to see Carlos. A crowd waited at the gate with horses and ox carts. Many arrived on foot. I set up the projector. We had no *cantina* and no one to take the entrance fee.

"Carlos. Help me out. Go next door and buy a litre of *caña*. While you're there, ask Antonio if he'll provide a *cantina* on his terms, but remind him it's for the school."

Carlos came back with Antonio, each with a case of beer,

and Carlos with a litre of *caña*. A cold breeze blew from the south. People sat huddled on the benches waiting for the movie to begin. I collected money from most of them. Carlos found a glass in my room and filled it with *caña*. He took a sip and handed it to me. I drank and put the glass on the floor and I asked Carlos to start the generator. When I heard the motor start I let the reels roll. A Charlie Chaplin film appeared on the wall of the *Centro de Salud* in front of a group of cold viewers. Antonio sold some beer. Carlos retrieved the glass from the floor and we shared it and another and another.

Half way through the reel, and the *caña*, I looked down and saw that the take up reel wasn't reeling the film. All the film that had shown had fallen directly to the floor in a spindly mess like a tape worm growing in volcano-shaped form. I grabbed Carlos and pointed. He stumbled off laughing. I recognized the humour but realized my responsibility, too. Some in the audience heard the laughter and turned and saw the film on the floor. I let the entire film run and started the next reel, cartoons, and I dragged the film into my room. I later rewound the entire reel by hand and then ran it through the projector to make sure it was tight and still functioned.

Carlos and I woke up to an empty litre of *caña*, beer bottles, a pile of film, a mound of crumpled guaraní notes, and heads as big as Brazil. We hauled water from the well to wash and tried to reconstruct the last hours of the night before. Carlos convinced me that no one would ever want to see a film at the *Centro de Salud* again.

We walked to the Medina house for breakfast and we waved to Antonio as we passed his house. I wondered if we owed him money, but I didn't want to ask, because I know he'd say that we did.

"How did the movie go?" said *Don* Icho.

"Great," I said, looking at Carlos.

"A good crowd," Carlos said.

Ña Buena served *mate*, and then fried eggs with *mandioca*. We ate and then told *Don* Icho that Carlos had to leave.

"Is it possible to take Carlos to his car by horse, *Don* Icho?" I said.

"That's no problem," he said. "Rodolfo," he called to his son, "Get us three horses ready."

We sat talking and Rodolfo led three horses to the house and prepared them with tack.

While Carlos stayed, I noticed that *Ña* Buena used a tablecloth and dishes I'd never seen. She dressed differently and so did Noemi, even when she washed clothes. Neighbours stopped by with weak excuses to see this other American with such charisma.

Don Icho, Carlos and I rode horses to the fence that marked the border of Guazú Cuá from the *estancia*. Carl dismounted, thanked *Don* Icho, climbed through the fence and walked to his car. We waited until he had started his car, beeped and pulled away.

"He's a good man," *Don* Icho said.

"He is."

"When do you think he'll come back?"

"I'll ask him when I'm in Asunción."

He never returned. I stayed at his apartment several times, but I believe he left Paraguay before me.

Radio *so'o* lived on Carlos for the next few days. *So'o* is meat in Guaraní, and when people stand in line waiting for a cow to be butchered they gossip and call it "radio *so'o*." The people loved him. His visit took me away from the reality of my site for a short time. Vilma and her group stood against me, and I spent too much time with Noemi, which everyone noticed before I did.

I found myself at a crossroad. I, a man who lived thirty-four years without having a serious commitment, now realized I was infatuated with the most unlikely woman. And that infatuation had turned part of the community against me. Notwithstanding, I found women without makeup, without dyed hair, without jewellery, without designer clothing, without vanity, exceptionally attractive, and Noemi was just that. I must speak seriously with her.

21

On September 12, 1989, a dreary, dark day of clouds and downpours, I moved into Noemi's house. Her entire family came to move me, but they did it late at night to avoid gossip. There would be less impact after the fact. We moved everything except my *ropero*. We splattered through mud and puddles and over and through fences, with my bed, table, pots, dishes, and clothing, behind the *Centro de Salud*, behind Antonio and Vilma's house, to the Medina house, and to my new room, laughing all the way.

I moved in to the Medina residence and I had some misgivings. Why do they want me there? It's certainly not for money, because they have helped me much more than I have helped them, and they seem to be self-sufficient and happy. Are they simply friendly folks protecting me from the evil in the community? Is it because of my relationship with Noemi? And what will the neighbourhood reaction be when women who leave their beds to milk cows by the light of dawn realize I have moved from the *Centro de Salud* to the Medina residence?

The Guazú Cuá part of the Medina clan consisted of Noemi, Augusto, Rodolfo, and Joel. Noemi had two married sisters. Estela, a year or so older, lived near Asunción and Ismenia, a year or so younger, lived in Ciudad del Este (which was called Puerto Presidente Stroessner before the coup) on the Brazilian border. Guido, a brother, lived with Estela. Francisco, younger than Rodolfo, lived in Paraguaraní with a family that owned a bakery. The next youngest brother, Guillermo, lived in Paraguarí with a family that owned a restaurant. Fabio, the oldest brother, served in the army.

Noemi herself had lived with a family in Sapucaí for some years and then with another in Asunción. *Don* Icho farmed out his children to work in the houses of the well-off so they could attend better schools and, in the case of the boys, learn a trade as well. Fatima lived with the Medina's for the same reason. She could attend school and earn her keep by helping out around the house and with the animals. I found the concept strange at first, sending one's children to live with and work for other families, but I saw the harsh wisdom, too. Although the children didn't see their parents often, they learnt responsibility and went to better schools than they would have at home. Some only returned to celebrate holidays and birthdays, or to attend funerals or weddings, and others, like Noemi, returned home because of a preference for the country life-style.

Almost everyone could read, write, and calculate. When I brought newspapers from Asunción, people read every page and then passed them on to be read again. Most had gone through the sixth grade, and most in a classroom with no electricity, in a school with no indoor plumbing. The teachers had the only books and they copied from the books onto the blackboard. The students copied from the board. Despite the literacy rate, education beyond the rudiments seemed to be reserved for the entitled children of Stroessner's men, military men, smugglers, and the corrupt. The Stroessner regime had no need for a nation of thinkers, and that would take generations to change.

I met all the Medina children when they came home to visit, with the exception of Ismenia. I also met some of the offspring of *Don* Icho from another woman in Potrero Arce. His children there had already formed families of their own. That toothless *Don* astounded me with his stories of the Chaco War of which he was a veteran.

Beside his participation in the Chaco War, *Don* Icho captivated my interest with his stories of his wanderings, such as rowing contraband to and from Brazil across the *Rio Parana*. *Don* Icho never cultivated crops, but he sold crops, transported crops, and provided services such as his dilapidated billiards table. He acquired land, too. He raised cattle, sheep, pigs, and chickens. Rodolfo cultivated some of the family land with cotton, corn, peanuts, and beans.

I came to admire this man, born in 1914, a father of many, an excellent horseman, in fine health for a man who hadn't been to a doctor ... ever? He knew where to find the medicinal herbs and how to prepare them. He cured me of bronchitis once with a concoction. Yet he didn't believe that the United States had landed a man on the moon. No man helped me more than *Don* Icho, and he did it as if I were his son.

But I still hadn't become accustomed to Paraguayan time. In the *campo* there are no hours. It's later, early, late, soon, in a little while, and these terms drove me nuts. I lived with the conditioned illusion of measured time, rather than time as now.

"What hour is early?" I'd ask. "What do you mean by late?"

It took time—plenty of it—but I reached a point where I knew that a point had a spread of a few hours. The same held true with directions or distances. "How far is it?"

"It's not far."

"But how far? How long will it take to walk?"

"It won't take long."

Jorge once told me he'd be at the house *tempranito* to build the *fogón*.

"Jorge, what time is *tempranito*?"

"Early."

"Jorge! What time is early? Come on. Is it five, six, or seven or what?"

"*Tranquilo*, Marquito. *Tempranito* is early. After we wake up, eat, and take care of business—is the time. Don't worry."

Don Icho and *Ña* Buena saddled horses and rode out before dawn to see his dying brother.

Jorge and I built the *fogón* and the chimney in a half-day. I hauled water from the well, mixed the mud in a pile, and handed the mortar and bricks to Jorge while I translated the directions (not that he needed them) and he constructed the *fogón*. He slapped the red mortar from his trowel and fitted a brick and continued along layer by layer until the *fogón* took shape. We filled the dead space with broken bricks, mud, sand, and chunks of roofing material. We tamped it solid and continued.

We installed the oven and the *plancha*, the iron sheet or the cooking surface. The *fogón* looked like an "L" that fell over with its short leg up. This is the part that contained the oven. The longer part of the "L" held the *plancha* and the fire box. The fire heated the *plancha* and the chimney drew the hot smoke from the fire box. Before it passed from the firebox to the chimney, the hot smoke was forced to circulate around the oven. It worked well and used much less fuel than an open fire.

Jorge and I erected the chimney from the *fogón*, outside and above the dry straw roof of the kitchen structure.

Above our heads, above the *fogón*, the underside of the straw roof almost dripped black, tar-like, creosote from years of smoke. Now the chimney channelled smoke out of the area, but I wondered about a spark from the chimney on the straw roof. Did I do the right thing? Am I unjustly paranoid? Did I not comprehend the reason for cooking on an open fire on the floor?

Don Icho and *Ña* Buena returned the next night and arranged to send Noemi to help with her dying uncle. She rode out well before dawn.

I built a *fogón* in the Medina house, the house in which I lived. I did it as an experiment and as a gift to the family. The neighbours saw it differently. I did it for Noemi.

I had orders for other *fogones*. Nidia, the most open-minded teacher, wanted a *fogón*, too. I waited for the parts to become available.

One morning I went to get a bucket of water from the well. The brick-lined, shallow well remained uncovered. Mud surrounded the well and the bucket and rope were on the ground, muddy. Chickens pecked around and a pig slept nearby. I squished my toes in the mud as I considered my morning routine. I lived contrary to what I had been sent to teach. Shoes aren't always practical. The locals live on a long-term camping trip.

I dropped the bucket with the muddy rope into the well and pulled up some water. I poured it into an aluminium basin to wash my face and hands after I used the outhouse. I wanted coffee instead of *mate*, so I walked over to Antonio's to buy a small packet. There Vilma confronted me.

"If you don't move out of that house no one will work with you again," she said. "And Noemi has a big friend who will surprise you one night. You don't know what you have started. I'm sending another telegram to Peace Corps and one to Dr N_____ to let them know about your deplorable behaviour," she snorted, storming off toward the school. Antonio grinned and shook his head.

I decided to go to Asunción to talk to Vita, and made a quick trip to the Peace Corps office. I didn't see Vita, but I spoke to her helper, Nanci. She said they'd be at my site on October 5, and she advised me to steer clear of Vilma until then. Nanci already knew the story. Vilma had sent messages and made phone calls. I spent the night on Luci and Carlos' couch and I returned to Guazú Cuá the following morning.

As I trudged into Guazú Cuá some people waved to me and others turned away. I knew there were factions within the community, but I didn't know who belonged to which faction. I slipped into my room and sat on my bed. *Ña* Buena strutted about yelling in Guaraní. I heard chickens squawk and dogs squeal as she hit them with a short whip. Joel cried in the next room. I wanted quiet. I'd been spoilt as an American when I could have my own space when I wanted it. I looked at all the families living in one or two-room houses and realized that privacy was as opulent as indoor plumbing, electricity and reliable transportation. Most days I enjoyed the lack of luxury, except when I squatted on planks over a hole in the ground holding my breath from the stench of human excrement, deafened by the drone of flies.

Noemi returned from her trip to her uncle's. He remained ill in bed, yet her return to the house lifted my spirit. I learnt that the next morning I'd ride a horse with Noemi and Fatima to Sapucaí. They'd return the same day, but would show me where I could leave the horse while I went about searching for parts to build *fogones*, and to see about a grant for the school kitchen. I also had to find a way of getting my pump to Guazú Cuá.

We gathered in the kitchen and drank *mate* and ate boiled eggs and *mandioca*. Fatima held Augusto. Fatima, a nine-year-old, already knew how to care for an infant or toddler; she could build a fire, cook, gather wood, pick cotton, milk cows, and ride a horse. She did well in school, too. I helped her with her homework, but I usually confused her because my own learning had been so different than hers.

Rodolfo prepared two horses for us, one for Noemi and Fatima and one for me. We mounted up and rode out at dawn on a clear, spring morning. We took the same road I had taken the first time I walked to Guazú Cuá, over Cerro Verde. Fatima, in her ill-fitting, non-matching clothes, sat smiling behind Noemi.

I now understood why people rode horses and not mountain bikes. Nothing looked foreign, not the ox carts, the rutted road, the straw-roofed houses or the barefooted people about their morning chores, milking cows, building cooking fires, and saddling horses. We crossed the *cerro* and descended into Sapucaí after two and a half hours on horseback.

We rode to a house with a large fenced yard. Noemi dismounted, passed through the gate and to the door and clapped. The door opened and a thin woman in a robe opened the door and exchanged kisses on the cheeks with Noemi. They spoke and laughed and gestured toward me. I watched women leading donkeys down a steep hill from *Cerro Rokê* along a trail and into town.

"Marcos, let's go," Fatima said, as she leapt from her horse. I slid off mine and she tied the reins of the two to the fence. We entered the house where the *señora* kissed each of us on both cheeks and told us her name. Noemi conversed with her in Guaraní too fast for me to follow, though their gestures and shared smiles suggested that they spoke of me.

Noemi turned to me and said that I could leave the horse there and I could sleep there if I needed to do so. She and

Fatima had to leave. She did say that she lived at the house for a couple of years when she attended school in Sapucaí. Noemi and Fatima said goodbye and left. The *señora* and I stood looking at one another. She spoke to me in Guaraní and I responded in kind, annihilating the language, to which she responded with laughter and then spoke to me in Spanish.

"I'll have a boy take care of your horse. When you come back, if you want to stay here, you are welcome. Your horse will be fine. I've known *Don* Icho for years."

A thin, dark teenager in shorts led my horse to the open field behind the house. He took off the saddle and gave the horse water and bran. I watched from the bus stop. I waited about an hour and then caught a bus to the terminal in Paraguarí. I walked two blocks from there to Route 1 and caught an express bus to Asunción.

I stayed with Carlos and Luci again, but I spent most of my time at the Peace Corps office. I had written a draft of a grant for the school kitchen and I had it proofread and then typed it and mailed it. My pump sat in *Proyecto Paraguarí*. I found some *fogón* parts in San Lorenzo, near Asunción. I bought them and arranged for Vita and Nanci to pick them up when they went to my site. I felt as though I had accomplished something, so I went to the Baviera to see if any members of my group were there. Many sat laughing around a table full of beer mugs. I heard that some had been in Asunción for weeks, waiting for roads to dry or waiting for supplies or money or simply waiting—mostly the latter.

I made my way to Luci and Carlos' apartment, slept on their couch, and left early enough to stop in Paraguarí to speak with Rosa about the pump and then to make it to Sapucaí, saddle my horse, and ride into Guazú Cuá as the first rain drops fell. I thanked the horse and again realized why the locals travelled by horse.

Rain fell all of that night of October 3, 1989. Rain fell in sheets and gusts and torrential downpours, saturating the ground and swelling the drainages. Vita and Nanci's visit looked bleak. They would visit and they would bring all my material, but it could be weeks before they could get the truck in to the site.

I bided my time reading, writing letters and studying Guaraní with Noemi. I found myself in a situation I would have considered most unlikely. I loved a woman with whom I did not share a common language, a common culture, and our educational differences spanned distances, although, from what I'd learnt in Guazú Cuá, the locals, despite my learning, knew much more than I about living in their country. I would have been lost without them. They could start a cooking fire in minutes where I spent an hour. Every man could build his own shelter with local materials, and a machete, a shovel, and pliers. I hadn't a clue. I may have had the book education, but I lacked the life skills.

I wanted to run to a doctor every time I felt ill, but the locals knew the herbal remedies that grew around them. I knew some history and philosophy, and some math and science, and had read hundreds of books, but I couldn't castrate a pig, or slaughter a steer, or break a horse, or prepare a field for planting, or dry meat, or drive oxen. Who is teaching whom here?

I found myself in love. Not only did I find myself in love, which I had experienced before, but so in love I'd considered doing things the Paraguayan way. The only way for Noemi is marriage because of her son, Augusto, who probably thinks he's my son anyway, and because anything less than that may leave her with a second fatherless child. I've considered it. Had I been in the campo too long? Did I see a diamond or zircon?

22

On Wednesday, October 4, 1989, I rode to Franco with *Don* Icho, *Ña* Buena, Noemi, Fatima, and Joel. We rode for three hours over hills, along trails, through stands of coconut trees, beside tall bamboo, and through forest to reach *Don* Icho's late brother's house, where Noemi had been helping just a short time ago.

After the burial—and he had been buried, customarily, within twenty-four hours before his body began to decompose—the family organized a *novena*, an event in which the rosary is said every day for nine days. At the house an altar held a photograph of the deceased, statues of saints, candles with black ribbons, and, in this case, the Paraguayan flag, because, like his brother, *Don* Icho, he served his country during the Chaco War.

The women of the family wore black and some of the men did, too. Others had black strips of cloth tied around their arms or sewed onto their shirt pockets.

Three dozen men, women and children attended the rosary the day we arrived. Everyone spoke Guaraní, and people wanted to know about me, but they wouldn't ask me. Instead, a person would ask Noemi or *Don* Icho.

"Who is he? Where is he from? Why is he here? Does he speak Guaraní? Does he like Paraguayan women? Does he drink *mate*?"

I understood enough to know when people talked about me. Their gestures helped. Although I felt uncomfortable,

I thought about how such behaviour was the norm in the Paraguayan *campo*, even though it might have been considered rude by some in my own culture.

After the rosary the men shared their *caña* with me, beneath the sparkling blanket of stars in the southern sky as we sat beside a stand of tall, waving bamboo. I conversed with a few gentlemen in a melding of Spanish infinitives, Guaraní, gestures, all facilitated by the tongue-loosening *caña*.

"Let's go to sleep," *Don* Icho said, tapping my shoulder and pointing to a nearby house. I shook hands all around, accepted a last slug of the *caña*, and followed *Don* Icho to a straw-roofed house a short walk from his late brother's house. A woman in black met us at the door and walked us to a room with a hard-packed red dirt floor, where a candle flickered on a chair between two cots. We undressed and hung our clothes from pegs on the *vigas* above the low wall of rough-cut planks. We each pulled up a cover and stretched out on our cots. I stared at the construction of the straw roof until *Don* Icho reached over and extinguished the flame of the candle between two fingers.

The next day the women prepared food all morning. The men sat talking, drinking *tereré* and waiting. I sat with the men but wished I'd brought something to read. The Guaraní exhausted me.

In the afternoon we all sat at tables beneath trees, grape arbours, and *galpóns*. We ate chicken—fresh killed that morning— in a tomato sauce with pasta and bread and *mandioca*. After the meal we men napped. At about 5:00 p.m. people began to arrive. An hour later fifty people gathered around the house of the deceased. We said the rosary, and the family provided coffee and snacks for the mourners. *Don* Icho prepared the horses while we drank coffee. We left that evening and rode for Guazú Cuá, one hour in the last light of day and two in the darkness.

During my absence from Guazú Cuá, Nanci and another volunteer, Leslie, visited my site. I had left a note just in case I missed them. They, in turn, left me a note.

They visited Vilma and said she wouldn't budge. She wanted me out of the Medina residence. She told Nanci that she'd sent a telegram to the Peace Corps office about me. Nanci knew, but said she hadn't heard about it yet. She and Leslie dropped off my pump and the *fogón* parts I'd left in Paraguarí. They left them at my house, the Medina residence, with Rodolfo.

We arrived in Guazú Cuá from Franco on Thursday night. The next morning Dr N_____ and his helpers raced into the community in a white, four-wheel drive Mercedes Benz. Dr N_____ sent for me and I walked over to the *Centro de Salud*.

"I've heard that you are not doing your job and that you are living with a family of which the community doesn't approve."

"With all respect, Dr N_____, I don't work for you and where I live is my business."

"If you're not doing your job you don't need to be here."

"Do you mean that if I'm not doing your job I don't need to be here, Dr N_____?"

"Let it go. I'll speak to Vita about your work and living situation."

"Great. I'll appreciate that."

***

On Saturday, October 14, 1989, I left Guazú Cuá on a trip to Caaguazu and Asunción. Before I went on the trip, Noemi and I spoke.

She told me Augusto was too young, she didn't go to fiestas, she was simply a *campesina*, she didn't know what her parents would think, we had a difference in languages, yet, finally, she said she, too, had the same feelings that I had expressed, but she was prepared not to do anything about it.

I admired her strength. I would not let it go at that. Noemi would. She accepted her lot in life. I would complicate her life or she mine. She had the ability to let go and to accept the present for what it was.

I walked out of Guazú Cuá that day thinking about what she had said. I walked for an hour before a truck hauling cotton seeds stopped for me, picked me up, and dropped me off at the main road. There I caught a ride in the *mixto* to Paraguarí, and, from there, a bus to San Lorenzo where I took Route II to Caaguazu, arriving at about 8:00 p.m. There I found David Lindsay, from my group, in his hotel room at his site.

David, a retired engineer, joined Peace Corps a year after the death of his wife. We had dinner together at the hotel and then I rented a room near his.

In the morning we ate at the hotel and then left to tour the market in search of *planchas* and *hornos*. These items were said to be more plentiful in Caaguaazu and less expensive than in other cities. This I found to be true, but I could only transport a couple on the bus. I had to change busses a few times, so logistics became an issue. Then I'd only be able to get the parts close to Guazú Cuá, and I'd have to find someone to haul them into the site in a truck or ox cart.

I left David before noon. He may have wanted more company, but I wanted to move my parts. I boarded a bus and hauled my parts to Piribebuy where I changed busses. Piribebuy was the third capital of Paraguay during the War of the Triple Alliance. I boarded a bus with my cargo and we rode down toward Paraguarí, with fields of cotton, corn, sugar cane, clear creeks and green, leafy trees growing out of red dirt. We descended into a rocky draw and passed Chololo, a series of pristine waterfalls and a hotel where many newlyweds often spent their first night.

When the bus rolled into Paraguarí, I carried my parts to the *Proyecto Paraguarí* and left them there for Rosa or someone to take to Guazú Cuá. I then caught a bus to Asunción and went to Carl and Luci's apartment, where I found them home and they welcomed me in, though an aura of tension precipitated upon the apartment.

***

I left the apartment early in the morning and caught a bus to Paraguarí, shopped, and then boarded another bus to Potrero Jara, from where I walked the fifteen kilometres to Guazú Cuá in three hours.

Vita would bring my pump and *fogón* parts that I left at the *Proyecto Paraguarí* when she came during the next week to speak with Vilma and some others. She said that she wanted to learn the reality of the situation in my site. Only the weather would change her plans for a site visit.

Monday evening, I sat beside the well and watched the *campo* brighten with the rise of the full moon, while the laughter of men shooting billiards and drinking came from the front of the house. I heard on my short wave radio that the Dow Jones Industrial Average dropped almost seven percent the day before, news that may have come from another planet. I had a glass in my hand and a few bottles of beer in a bucket cooling in the well. I didn't want to join the young revellers at the billiards table, and I didn't want to sit in my room. During the day I spoke to Gladys, a woman with a house in the centre of Guazú Cuá, about a *fogón*. She wanted one. Jorge agreed to build it with me. Others wanted *fogones*, but I couldn't get the parts without taking a trip, as I did when I went to Caaguazu.

"Marcos," Jorge had said, "let's go to Sapucaí to the shop at the train station. It has everything to make *planchas* and *hornos*. I can go Wednesday."

"Okay. If you think we can get them made there."

"I know they have made them, and they have the materials in the scrap heaps. It's a matter of convincing them that they can make a little extra on the side and help the people in the campo at the same time."

"Thanks, Jorge. I'll be ready to go," I said. It would mean work for Jorge and would be a benefit to the community if we found a source for the parts we needed.

Noemi emerged from the darkness into the light of the full moon and she sat on the edge of the well. We chatted for an hour and she invited me to accompany her and Augusto to lunch the next day at Augusto's God-parents' house. I accepted. She left and I finished my beer while staring at the moon. It was actually the eve of the full moon, but I didn't realize it until the next night.

23

Vita visited my site on Tuesday, October 17. She brought my pump and the *fogón* parts. She visited Vilma and spoke with several other families. I waited at my house. When Vita returned she said that, even though Vilma remained stubborn, the community understood that I could live where I chose, communicate with whom I wished, and that I was in Guazú Cuá for the good of the community and the residents should take advantage of what I had to offer.

Vilma had interjected that I was in love with a less-than-desirable woman, and Vita told her that I was free to form relationships as I saw fit and it was no business of anyone but me and any other party involved. I imagined Vilma hyperventilating.

Vita left and I inspected the pump. Jorge dropped by and we planned the installation. Vilma remained a barrier, but she would have a difficult time convincing the parents that the school didn't need the pump.

In the morning I drank *mate* while I waited for Jorge and Julia. Julia arrived on horseback with Jorge walking beside her. *Don* Icho had prepared a horse for me to use. Jorge wanted to use my bicycle.

We three set out for Sapucaí, across the *campo comunal*, into a draw and across a rocky creek, past the brick administration building for *Colonia Santa Isabel*, the leper colony, and up and over the red dirt and the florid green of Cerro Verde, and then down across pasture and into Sapucaí.

We went first to the shop at the train station. The administrators had a meeting and we couldn't get in to speak to anyone.

"Come back tomorrow, or the day after tomorrow," a secretary said.

I bought meat and we rode back to Guazú Cuá.

***

On Sunday, October 22, 1989, Noemi and I talked about marriage. I proposed and she accepted, but with worries about Augusto. He already thought I was his father, since I was always around, and I talked of going to the embassy and to the Peace Corps office to get permission and paperwork. We talked about April as a date, but then decided we should do it sooner. We settled on the eve of my birthday, December 7.

This gave me a lot to do. I had to acquire papers from the United States to prove I was a confirmed Catholic (not really important for the Justice of the Peace), and other papers, such as a police record, or lack of one. I wrote to my sister and anyone who could help me get the papers, always wondering if marriage was the right thing to do, considering the length of time we'd known one another, language, culture, customs, and that I'd have a ready-made family. That's how my father responded. He asked if I really needed a ready-made family, and did I know what I was doing. I wondered myself, as did members of my group, Areguá 1. Wonder aside, I had made a decision and the blossom of love conquered doubts.

***

Vita told me I could buy a horse and get reimbursed from Peace Corps. I took an advance of $100 from my vacation pay, bought a carrot-red horse from *Don* Icho for seventy-five dollars, and then recovered my money from the Peace Corps. *Don* Icho loaned me the tack when he sold me the horse, and he wanted to buy the horse back if and when I left Paraguay. *Don* Icho had given the horse a Guaraní name, which I couldn't pronounce, so I renamed him Gulliver.

I rode my Peace Corps bicycle to the *Proyecto Paraguarí* and left it there for Vita to pick up and take back to Asunción on one of her trips. The bike failed in my site because the composition of the soil, when moist, stuck to the tyres. And with my carrot-red steed, Gulliver, I didn't need a mountain bike.

The first week of November, mid-spring, I wrote and sent twenty-seven letters announcing my proposed wedding. I knew not one of the recipients would travel to South America to attend the event, yet, without shame, I hoped some might send a few dollars.

Meanwhile I waited for the documents I needed to marry, such as my certificates proving that I had been baptized and confirmed a Catholic, and hoped those documents existed. My sister, Fran, in New York, did the foot work. I also had to wait two weeks for clearance from the American Embassy to marry. It cleared on November 12, 1989.

Thursday, November 16, I rode Gulliver to Sapucaí to the train machine shop. I met with the director after being sent from one person to another for three hours. I submitted a request for thirty *hornos* and *planchas*.

"*Don* Marcos, we will fabricate the parts you need, and, on behalf of the people, I congratulate you on your mission to help my country and the people of the *campo*. We are happy to help you in your endeavour. Return in a week and we'll have the first several sets completed for you," the manager said, standing behind his dusty desk and extending his hand to me. We shook hands and he walked me out of the building with a hand on my shoulder.

I left happy and mounted Gulliver. He knew I wasn't a Paraguayan, maybe because I changed his name, but more likely he knew a horseman when one sat astride him. He twisted his neck back to bite me and paid little attention to the bit in his mouth. Gulliver wanted to run home and that's what he did, with me pulling on the reins and finally hugging his neck and grasping his mane while I slid one way and another on his back. As we loped into Guazú Cuá I slowed Gulliver to a trot, and then a hesitant walk. We stopped at the local sawmill, really a hand and chain saw operation. I had ordered some wood ten days earlier to make forms for *losas*, or concrete slabs for outhouse floors, and my order should have been ready.

"It's more work than you think, *Don* Marcos. We'll have it done next week, but we can't do it for 3,000 guaranies a form. We have to charge 5,000."

It figures, I thought after I agreed on the price and grabbed the reins of my steed. Gulliver turned in circles and snapped at me as I tried to get a toe into a stirrup. I leapt onto his back and he reared up and raced toward the house. My legs hung on one side of the saddle and my head on the other. I worked my way onto the saddle and caught the reins just as Gulliver slowed to a trot to enter the yard of the Medina residence.

Noemi asked me if I'd like to go to the arroyo to wash the horses and the saddle blankets. I did. I wanted to and I went, but I failed to realize it was a family event. We weren't going to be alone. We were all going to wash horses. And it would take hours.

The next day the doctors from GTZ came to Guazú Cuá and asked the people about problems with the *Centro de Salud*: Were the people happy? What things could change?

No one said anything.

"If a health professional stayed here it would be more effective," I said.

"But the records show that about forty people a month visit the *Centro de Salud*."

"If a health professional stayed here hundreds would use the *Centro de Salud* every month. As it is, people come here for aspirins or vitamins. That's all," I said.

I left the meeting defeated. I began to understand cultural differences, and I planned to marry into such differences. I found a perfect bride. She wore no cosmetics. She could wring a chicken's neck, pluck it, clean it and cook it over an open fire of wood that she, herself, chopped. I never heard her complain about her lot in life. God had blessed her (or cursed her) with the gift of caring for others, especially babies, the elderly and the terminally ill. Her patience, softness, ruggedness, beauty and her unrelenting love of family and life grabbed me, yet I wondered how we would get on intellectually. And I wondered if a formal education had anything to do with our relationship.

***

Women served men. Men worked in the fields and women, with the exception of harvest, worked at home. Women cared for livestock, gathered firewood, cooked, cleaned, hauled water, cared for children, and, after all, waited until the men ate before eating. This was not true in every household, but men receive preferred treatment in Paraguay. I, too, benefited from this behaviour, but not because

I wanted it. Such conduct was the norm. I hadn't become accustomed to being special.

I dismissed my doubts and we continued to plan a civil ceremony in Sapucaí for December 7, 1989.

I went to Asunción to the embassy to ask about what, exactly, I'd have to do to obtain visas for Noemi and Augusto. While in Asunción, I learnt that we volunteers would get a raise in our stipends. I would get 139,340 guaranies, up from 126,600 guaranies. With twenty-nine percent inflation, the raise meant little.

While in Asunción, I learnt that my supervisor, Vita, resigned on November 14, 1989. She hit a woman and a child in Paraguarí while driving a truck owned by Peace Corps. The victims suffered minor bruises, but Vita held a license that had expired more than a year earlier. She'd used government vehicles without a valid license.

On Thursday, November 30, 1989, Noemi's folks left for the settlement of Durazno with Fatima, a day-long trip on horseback and on busses, because Fatima's father was sick. They didn't know when they'd return and didn't know if Fatima would return.

On Monday, December 4, I rode to Potrero Arce with Noemi and her young cousin to get her maternal grandmother. We rode over the hill and to the south and west through woods and across fields in the warmth of the morning. Men and women walked or rode to the fields on horses or in ox carts. Brahmas grazed in open pasture and we spotted a group of grey rheas in the distance. We rode into Potrero Arce in the late morning. We drank *tereré* and ate fried *mandioca* with eggs while Noemi's grandmother gathered her bags. She would ride with us to Guazú Cuá and then would travel from there to Durazno to see her son, Fatima's father.

The four of us rode to Guazú Cuá and arrived by mid-afternoon in the heat. Noemi's cousin took the woman to another relative's house where she would stay until she travelled.

Tuesday we saddled the horses and rode out of Guazú Cuá, over Cerro Verde and into Sapucaí. We stopped first at the train machine shop to ask about my *fogones*.

"We haven't had time, *Don* Marcos, but give us a few more days and we'll have them finished for you."

We stopped at the Justice of the Peace to make sure we had all the documents we needed to have a civil ceremony. We tied our horses to a post in the dusty street in front of the blue plastered building with a mouldy tile roof. A circular sign identified it as a government building. We entered a room with a high ceiling from which hung a squeaky fan with slowly turning blades. All the tall windows were open, yet the smell of sweat, dust, horse manure and tobacco smoke permeated the room. We sat on a hardwood bench. Several others waited and all stared at me, the foreigner. Within an hour our turn came and we spoke to the Justice. He checked our documents and explained the procedures to us.

We made an appointment for Thursday morning, December 7, thanked the Justice, and walked our horses to a nearby restaurant, actually the front room of a house with four tables, where we lunched on the only thing on the menu. We ate a chunk of tough, fried beef with sautéed onions and a fried egg on top and with rock hard rolls on the side. I washed mine down with a beer and Noemi drank a soda. Other customers stared at us, the Paraguayan woman with the *Norte Americano*.

We left the restaurant and rode to the post office where I gambled and left some letters, betting on the chance that they would reach their destination. We made one more stop at a house where we planned to leave our horses after the civil ceremony so we could take a bus to Asunción and then to Ciudad Del Este on the Brazilian border. We obtained permission to leave the horses on Thursday and we rode back over Cerro Verde and down into Guazú Cuá in the afternoon heat. We unsaddled the horses and turned them out to the common pasture and then sat beneath the canopy of a *Paraiso* and drank *tereré*.

Noemi had planted three *paraisos* in a row, now tall with broad canopies, along the fence line of the yard. Many sat beneath their shade to share *tereré*, a respite from the sauna-like summer heat. Legend had it that one who planted the *paraiso* would be destined to travel faraway. We spent countless hours in the shade of the *paraisos* talking, drinking *tereré*, napping, eating, and simply enjoying the natural relief from the oppressive temperature.

24

Don Icho prepared the horses and he rode with us to the Justice of the Peace in Sapucaí on the hot, humid morning of Thursday, December 7, 1989. We tied the horses in front of the blue building and entered. A few witnesses from Guazú Cuá waited for us inside. Noemi stepped into the bathroom and changed into a dress she had made. An aide ushered us out of the waiting area, across a patio and into an office with bars on the windows. The Justice of the Peace sat behind a worn desk with a dusty calendar on the wall behind him. He spread some papers before us and we took turns sitting in the one chair to sign. I gave my camera to one of the witnesses to record the event.

The Justice completed a marriage certificate form by hand and affixed several stamps to it and then slammed his seal over the stamps. We asked for extra copies and he repeated the procedure and made four originals, none of which looked quite the same. We kissed, shook hands all around, and I paid the Justice and then paid again for each of the certificates and again for a family booklet in which we were to record dates of birth and other notable information pertaining to our marriage. I looked at the dilapidated room with bars on the window and at the witnesses with their broad-brimmed hats in their hands, and at the certificates and felt as if I'd fallen in the rabbit hole. I'd married in an antique aura of the surreal. The aide ushered us out of the office, across the patio, through the waiting area, and to the dusty street where our horses waited.

We mounted and rode to the house where we were to leave the horses. We had the woman of the house snap some pictures of us before we left. She had never used a camera before and didn't get any good photographs. Our witness didn't do so well with the camera, either. *Don* Icho turned our horses out to graze and stored the tack in a shed. He hugged Noemi and shook my hand and mounted his horse and trotted toward Cerro Verde.

Noemi and I waited in the shade until we heard a bus rumbling toward us, so we crossed the street and signalled to the driver. The bus stopped in a cloud of red dust and we barely had a foothold before it roared off again spewing a plume of diesel smoke. We switched busses a half hour later in Paraguarí and rode to Asunción. We went to the *Itapúa Residencial* and rented a room. I'd always gone there alone and I introduced Noemi as my wife to the owners and workers, but I'm not sure they believed me.

We went to the Peace Corps office to get my mail. I hadn't told anyone that I planned to marry, except for the Country Director, from whom I'd had to obtain approval. Surely word got out. We didn't see anyone from Areguá-1. We caught a bus to downtown and ate dinner and then returned to the hotel.

The next day we celebrated my birthday by wandering around Asunción, visiting Areguá, and resting at the hotel. December 8 is a national holiday in Paraguay, the Day of the *Virgin de Caacupé*, or Our Lady of Caacupé, *Nuestra Señora de Caacupé*. The City of Caacupé sits east of Asunción in the hills, or *cordillera*, in the *Departmento Cordillera*, and is the site of a formidable basilica. On December 8 each year people from all parts of Paraguay travel to Caacupé to worship. Many walk or ride ox carts. Most walk only the last fifteen kilometres or so. Several miracles occurred in Caacupé that are attributed to the Virgin.

The first miracle happened in the sixteenth century to an indigenous man who had been converted to Catholicism. He found himself surrounded by members of a hostile tribe or Spaniards and he hid in a massive tree trunk. He prayed to the Virgin Mary and promised her that if she saved him he would carve a statue from the tree trunk for her. The hostile group passed without seeing him, so he carved a statue. The area flooded and flood waters carried away most things including the statue. When the flood waters receded, the statue reappeared. This statue became that which is now housed in the basilica in Caacupé.

After a quiet day, we returned to the hotel. Early the next morning we left for Ciudad Del Este on the Brazilian border. It began as a relatively small town in the late 1950s and grew rapidly during the construction of Itaipu Hydroelectric Dam on the Parana River in the late 1960s. Brazil, Paraguay, and Argentina share a common corner near Ciudad Del Este. Across the Parana River is the City of Foz, Brazil and *Foz do Iguazu*, a majestic waterfall on the Parana River which may be seen from the Brazilian or the Argentine sides of the Parana.

Ciudad Del Este differs from other cities in Paraguay because of the trade and the people attracted by the trade. Asians and Arabs share the city with Paraguayans, Brazilians, and Argentines. The Tri-Border area is a key smuggling point as well as a transit point for many legal goods. Vendors crowd the streets. Almost anything can be purchased from fine clothing and jewellery to counterfeit name-brand electronics, sex and illegal drugs.

Some members of the Gonzalez family from Guazú Cuá (the first family with whom I stayed) relocated in Ciudad Del Este and worked as currency changers and street vendors. We planned to stay at their house for a night or two before returning to Guazú Cuá.

***

We dozed and watched the scenery as we rolled along Route 2 travelling east from Asunción. We passed Caacupe where so many had gathered and still clogged the road with vehicles, horses, ox carts, bicycles, and people on foot, and then along the green *cordillera* to Coronel Oviedo, the red earth of Caaguazu, and on to Ciudad Del Este. We had directions to the street sales table of a Gonzalez. I'd never met him, but Noemi knew him. We pushed through crowds and I kept one hand in my pocket where I kept my wallet and passport. Noemi spotted his table, about the size of a cafeteria tray beneath an umbrella. Counterfeit Rolex watches covered the table. They spoke in Guaraní and he gestured toward a hill. We shook hands and then Noemi and I left for the house.

We shoved against the current of vendors and potential buyers until we reached a road that left the main street. We looked at it, steep and muddy, and we heard yelling. We turned to see two Paraguayan policemen arguing with a black man. One had his pistol out. He punched the man in the stomach with the barrel. The man doubled over and yelled. The policeman put the pistol next to the man's ear and fired into the air. The man fell to the side from the blast next to his ear. I stood transfixed and Noemi grabbed me and we left the main road and climbed the slippery hill. I looked back and saw the policemen laughing as they walked away from the man.

We climbed to a house no vehicle could reach. Many lived there and our host offered us a place on the floor. We stayed that night, crossed into Foz the next day, and decided to leave that night. We did buy a sack of inexpensive merchandise in Brazil, such as socks, underwear, thread, and such useful items at a price much less than what we would have paid in Paraguay.

We smuggled the merchandise into Paraguay, which we didn't realize until we'd passed through customs. We walked to the Brazilian side, but we took a taxi back and agents waved us through. Had we been stopped, we would have paid a tax and might have had something confiscated.

We caught another cab to the bus terminal and bought tickets to Asunción. We decided to spend another night in Asunción instead of Ciudad Del Este. That night, though, was spent on the bus as an unlikely couple, man and wife, asleep and speeding along Route 2 West in an express bus.

***

On Wednesday, December 13, I walked from Guazú Cuá to the bus stop at Potrero Jara. I walked past fields and houses I now knew. I waved as I passed houses while I lost myself in thought. I trudged up and down hills in the middle of South America as a Peace Corps volunteer and now a husband to a local woman and a father to a son who didn't know his biological father.

I reached the house at the bus stop and went to sit in the shade and to drink *tereré* with the others who waited for the bus.

"Did you marry a woman from Guazú Cuá?" the owner of the house asked.

"How did you know?" I said.

"People talk," she smiled.

The news spread. Not that it mattered, but everyone I met had the same question. Meanwhile, I was steeped in cultural differences that I believed I knew, but I did not. I needed privacy and the Paraguayans in the *campo* had little.

Families lived together in one room. As long as I lived in Guazú Cuá, I would not have what I considered privacy. I had to learn to live with a family I'd married into but did not fully understand yet. I thought since I lived there, I should learn the customs and not consider the differences.

I caught a bus to Paraguarí, made some purchases, and caught a bus back to Potrero Jara. I hiked back to Guazú Cuá with a full pack. I dumped my pack onto the billiards table and went to draw water from the well to wash. I bathed and slipped into bed beside Noemi.

At 6:00 a.m. I awoke and Noemi had gone. I found her squatting on a low stool in the corral and helping her mother milk cows. I'd thought that the work would end with our marriage, but it continued. Noemi and her mother built a fire, boiled the milk, and cooked breakfast. I watched in the wet, cloudy morning, and I ate the eggs and *mandioca* they cooked and drank the boiled milk.

Rain fell all day and with the heat no place remained without the heavy air of humidity. One could be in the rain or out of the rain and be wet. What I did not have was a place of my own, which didn't bother Noemi. I would later discover that to be one of her strengths, but I struggled to live without my own space, especially with my wife.

25

The wind brought the smell of rain on a hot, clammy Christmas Eve day. We baked carrot cake in the *fogón* oven in the morning and then began to cut melons, pineapples, oranges, and bananas. We threw all the cut fruit into a five-gallon bucket and added grapes, six litres of red wine, a few litres of *sidra* (carbonated hard apple cider), and a half-litre of whiskey. *Clerico*, similar to Sangria, is the traditional Christmas drink in the *campo* and each household produces a batch with many variations on the contents. As the level of a container of *clerico* drops, someone raises the level with more wine or *sidra*.

Rodolfo arrived with freshly cut branches, straw, and coconut tree flowers. With these the family constructed a *pesebre*, a nativity scene. *Ña* Buena dusted a collection of brightly collared clay figures for the scene, which included the baby Jesus, camels, kings, sheep, Mary, and Joseph. Each dwelling in Guazú Cuá made *clerico* and built a *pesebre*, some on a table-top and some on the ground and quite spacious.

Don Icho packed the *tata kua*, the beehive oven, with *leña*, firewood, and set it ablaze. Smoke rose from every *tata kua* in the community. *Tata*, in Guaraní, means fire, and *kua*, means place of (as in Guazú Cuá—the place of the deer). Francisco, Willy and Guido were home for the holidays. They clamped a hand grinder to the edge of a table beneath the shade of a straw roof and took turns grinding corn.

Estela, her husband, Abraham, and their two children rested, after the trip from Asunción, in the shade of the *paraisos*, on home-made cots with thin, hand-cut strips of leather woven within a frame creating a hammock-like but stiffer surface on which to lie. Abraham drank *tereré* and Estela brushed her waist-length black hair. The children dozed.

Noemi and *Ña* Buena unwrapped chunks of home-made cheese and they cracked eggs and cut onions to use with the ground corn to make *sopa paraguaya*. People saved eggs and cheese and corn before Christmas, so not an egg could be bought during the days leading up to Christmas, except for an inflated price.

Black storms with booms of thunder and flashes of lightning passed in the distance but did not hit Guazú Cuá, yet the scent of rain hung like an aura. A group played volleyball at the Gonzalez house across from us, the house of the parents of Augusto's biological father. A boom box blasted disco music.

People wandered from house to house. At the Medina residence, a group would arrive and ask about me in Guaraní, but they would not ask me directly. They'd question and gesture toward me. I understood and wondered why they didn't ask me. Generally they wanted to know if I drank *tereré*, if I liked Paraguay, if I ate *sopa paraguaya*, and why I left my own country. They looked at me as they questioned, which reminded me that no other foreigner had lived in the area.

I thought it rude, but remembered that they may have been well within their rights in their culture. When they left they told others what they learnt about me yet lent toward hyperbole which bloomed like mushrooms after a rain.

The boys finished grinding the corn and *Don* Icho pulled

the coals out of the *tata kua*. Noemi and *Ña* Buena mixed the corn meal, chopped onions, crumbled cheese, battered eggs, and lard in baking pans. They put a pan in the *tata kua* and covered the opening so the heat would not escape. The fire heated the bricks and the heat would bake a few pans of *sopa paraguaya* before the oven cooled.

The women baked for hours. The men gathered in shady places to drink *tereré*. Some started hitting the *clerico*. Noemi, Estela and *Ña* Buena mixed *mandioca* flour, eggs, milk, and cheese for *chipa*. They rolled the *chipa* dough and twisted it into pretzel-like shapes, but as thick as bagels, and then baked the *chipa* alongside the pans of *sopa paraguaya*.

I attempted to help the women but they ran me off. The brothers worked at chores. *Don* Icho and I remained free. He rested beneath a *paraiso* and I picked up a novel, *Resurrection*, by Leo Tolstoy. I could barely read in the suffocating air, as a procession of white, grey, and black clouds passed overhead changing the light.

Don Icho arose and spoke to his sons. They built a pit and a fire to cook beef on stakes. I saw Noemi peeling *mandioca*. By afternoon we had a table piled with pans of yellow, greasy *sopa paraguaya*, soft, warm *chipa*, and chunks of well-done, tough beef, and *mandioca*. *Ña* Buena fitted a second table with a chequered table cloth and she set a punch bowl filled with *clerico* in the centre. She placed plates, glasses, knives and forks, and cloth napkins around the table. We loaded our plates and glasses and sat at the table or on a bed or a chair in the shade of the *paraisos*. We hardly used the knives or forks. We gnawed the beef from its bones and used hands to grab *mandioca* or *chipa* or *sopa paraguaya*, and washed the food down with *clerico*.

We ate and drank until our bellies filled and our eyes

drooped, so each retired to the least sultry place available to sleep, beneath the canopy of a *paraiso*, under the straw roof of a *galpón*, or inside a room with the doors and windows open. I awoke drenched from sweat and the humidity. I walked outside to the well and drew water to bathe. I filled most of a five-gallon bucket and carried it to the *bañadera* where I poured it over my head with a can until I used it all.

Afterwards, I sat with the family around the bucket of *clerico*. We sipped the concoction and ate the alcohol saturated fruit. We had no ice, but it stayed as cool as could be in the darkest shade. A table sat piled high with meat and *chipa* and *sopa paraguaya* covered with cloth to keep the flies off the food.

Flies buzzed and the odour of lard, smoke and burnt fat hung in the air, even with a breeze that smelt of rain, but the rain never arrived. Guazú Cuá remained dry.

At dusk we made a pilgrimage to a dozen or more houses. *Ña* Buena and *Don* Icho stayed behind to welcome guests. We trudged across fields until we arrived at a house. We clapped our hands to alert the family of our arrival and then we entered the property upon invitation.

"Come in, come in."

"Thank you, *Señora*."

"Sit here, by our *pesebre*."

"It's so beautiful."

"Thank you. We spent a lot of time on it."

"It came out well."

"Will you drink our *clerico*?"

"Of course. Thank you."

Señora Luisa served each of us a glass of *clerico* from a large glass bowl on a table near the *pesebre*. She served a glass to her husband, *Don* Juan, too. We sat in the waning light with dark clouds passing overhead and sipped our *clerico* and munched on the fruit.

"Your *clerico* is excellent, *Ña* Luisa."

"*Gracias.*"

"Your *pesebre*, too, is very nice. How did you make that scene? It's so colourful with the fresh coconut blossoms."

"*Don* Juan and Antonio cut them today. We're glad you like them."

"Thanks for having us. We're going to go to Nidia's house now." We handed *Ña* Luisa her empty glasses and we all shook hands with *Don* Juan before we left for another house. We climbed through the barbed-wire fence and walked toward the glow of a lantern at Nidia's house.

Another family passed us along the way and we stopped and wished one another a Merry Christmas. We all felt good. At Nidia's we clapped and she invited us in. We held the barbed-wire for one another and went to the *galpón* where the lantern shone. Nidia greeted us and we entered and shook hands with her lanky husband Timoteo. Their daughter, Zuni, a smiling, big-eyed, long-haired, dark-skinned teenager, busied herself by filling glasses of *clerico*, which she served to each of us. We sat near their *pesebre* populated with doll-sized figures of kings, sheep, camels, Mary, Joseph and the baby Jesus in his straw-filled box, and we drank our *clerico*. We scooped the last bits of alcohol-drenched fruit from our glasses, thanked the family and walked to the Caceres' home.

In the humid, cloudy darkness we saw the red tips of cigarettes, candles, and flashlights moving between lit houses. We trooped along single file across drainages, across fields, and toward the light of *Don* Caceres' house. We found him sitting outside with one of his blue, green, and yellow parrots on his shoulder and beside his *pesebre*. His wife and a daughter prepared the meal which they would eat at midnight.

"Amelia! Guests," yelled *Don* Caceres.

Amelia, a green-eyed teenager appeared barefoot with a pitcher of *clerico*, a ladle, and a tray of glasses. She served each of us, exchanging words in Guaraní with everyone, until she reached me.

"*Don* Marcos," she said as she handed me a glass and ladled it full of a very fruity and carbonated *clerico*.

"*Gracias. Muy amable*, Amelia," I said. I toasted *Don* Caceres. We clanked glasses and drank and talked and admired the *pesebre*. Another parrot, identical to the one cocking its head from one side to another on *Don* Caceres' shoulder, watched us from a vine.

"Stay and eat with us," *Don* Caceres said.

"We'd love to, but we have a meal waiting at home," Noemi said.

We finished the *clerico* and, after visiting several houses, found our way home to eat the second Christmas meal. In our absence a dozen families stopped by to see our *pesebre* and to try our *clerico*. *Don* Icho continually cut fruit and poured wine, hard apple cider, and soda into the *clerico* bucket to replenish the supply.

Ña Buena spread an embroidered cloth over the table in the open air beneath the straw roof where she piled baskets of *chipa* and *sopa paraguaya*, plates of beef, plates of *mandioca*, salt dispensers, a bowl of lettuce and tomatoes, and glasses.

We sat and filled our plates. I didn't feel like eating after so much alcohol-soaked fruit, but I did, and sat moist from the wet air, and enjoyed partaking in the local custom.

After eating we sat under the open sky and watched the clouds and fire flies. I immersed myself in family, yet this sultry early morning of December 25 did not fill me with the spirit of Christmas, despite the tour of *pesebres*. We exchanged no gifts. Paraguayans, at least in the *campo*, exchange gifts on January 6, the *Día de Los Reyes*, or the Day of the Kings. Gifts aren't exchanged, but children leave their shoes outside a door or on a windowsill and, in the morning, find that the Kings have left gifts in their shoes.

Everyone found places to sleep, outside and inside. Noemi, Augusto, and I slept on our bed inside and beneath a mosquito net. We awoke within a few hours at dawn to begin Christmas day. *Ña* Buena already had water boiling on the morning fire, and she had buckets of water ready for the family to use to wash. She must not have slept.

The neighbours, the Gonzalez family, set up a volley ball net in the street. Guitar and harp music and songs in Guaraní blared from radios or cassette players in many houses. Men who hadn't made it home the night before stopped by for a morning glass of *clerico*.

I saw nothing to promote buying gifts in the days leading up to Christmas, but in the markets in Paraguarí and Sapucaí, I saw more fruit at higher prices and more clay figurines of animals and religious figures. Vendors sold the flower of the coconut tree for *pesebres* or for Christmas decorations. Cases of *sidra* from Argentina to use in *clerico* filled stalls in the market.

Since the price of food increased before Christmas, people like *Ña* Buena hoarded eggs from her own chickens. She also made her own cheese, her family butchered its own cattle, and she made her own *mandioca* flour and ground her own corn meal from her son's crop. She traded for the fruit for *clerico* and whatever else she needed.

I didn't miss the incessant advertising of the Christmas season that I'd known in the United States, but the humid summer days did not have the feel of the celebration of the birth of Christ. I'd always associated the holiday with winter. Only the nativity scenes suggested Christmas.

Here people danced in the road, played volleyball, drank *clerico*, ate much more food than usual, and wandered about to approve of one another's *pesebres*, without a commercial Christ in the world of consumerism. Figures of Christ appeared in every *pesebre* in Guazú Cuá, and extended families broke bread together.

26

I spent the early morning hours strolling in the common pasture with a length of rope hidden behind my back. The horse wouldn't move. Gulliver grazed and watched me. When I got close enough to throw my rope, he bolted and stopped to graze again about 100 feet away. I walked to him again and kept my rope hidden. I walked right up to him and stroked his nose and neck with my left hand, while I held the rope with the other. I brought my right arm around to get the rope on him and he snorted and galloped off a hundred feet or so. I felt the eyes of many watching and I could have had Rodolfo catch the horse for me, but I wanted to do it myself. I followed Gulliver around the common pasture for over an hour, and then Rodolfo joined me.

"Give me the rope, Marcos," he said.

I handed him the rope. He walked straight to Gulliver, said something in Guaraní, tossed the rope over the horse's neck, tied the rope and led Gulliver to me.

"Thanks, Rodolfo," I said. "I need to learn to catch the horse myself."

"He knows you're not a Paraguayan," he laughed.

I rode Gulliver to Sapucaí that morning of Thursday, December 28, in the hot, thick air. We trotted out of Guazú Cuá and walked up and over Cerro Verde, passing men behind ploughs pulled by oxen and smouldering piles of brush and branches from cleared land. The canopies of trees shaded most of the rutted, rocky road. I waved to families in their straw-roofed houses. Everyone knew me by name, although I knew few of the people who greeted me.

We crossed the more populated crest of Cerro Verde and dropped toward Sapucaí, still some several kilometres distant. I planned to visit the judge to get some papers concerning our marriage, and I wanted to speak with someone about my *fogón* parts at the machine shop at the train station.

"*Don* Marcos, we are in the holidays," the manager said, his shoulders shrugged and his arms outstretched, hands palms up. "We will have your order by January 15. It's impossible before because almost everyone is off until then," he said with a last shrug and hand gesture.

"But you said they would be ready before this, before anyone left on vacation."

"*Don* Marcos, you know how it is. We get busy and something is left behind. We have to keep the train running first," he said, shaking his head.

I climbed on Gulliver and turned him toward the public building in Sapucaí. Sweat dripped down my back and my chest as Gulliver kicked up red dust with each hurried step. I found the building closed with no sign to explain its closure.

"Why is this building closed?" I said to three men. They sat on stools drinking *tereré* in the shade.

"You might find someone here next week, *Don*, but the judge won't work again until February or maybe earlier. It's the holidays for some," one of the men said as he held a *guampa* of *tereré* up to share with me.

"*Gracias, pero, no, gracias*," I said motioning with the palm of my hand. "I need to eat first," I said.

"*Suerte, Don*," the man said.

I clicked my tongue and dug a heel into Gulliver's side and we started along the road towards Cerro Verde. I stopped and bought enough *chipa* for the family from a woman who always sold at a bus stop. Hers had more milk and cheese than most, which, I supposed, made it softer and tastier, but she priced hers higher. I liked it more than any other I had tried, yet the people in Guazú Cuá said she charged too much, as they wiped crumbs from their mouths of the *chipa* I purchased in Sapucaí.

Gulliver ran most of the way to Guazú Cuá, with me bouncing in the saddle and trying to remain upright, yanking on the reins, as if I controlled the horse. I waved to the farmers and their families while some gestured to me to stop and drink *tereré*.

"Hey *rubio* (blonde), come home with us!" yelled a couple of young women in an ox-drawn cart.

I smiled and waved as Gulliver pushed on up and over Cerro Verde and down into Guazú Cuá, where he wanted to graze without a saddle or me on his back. He acted quite differently with Paraguayans. I loved him anyway, despite his faults, including that he often attempted to throw me.

On Saturday, December 30, I took the sucker rod from the school well pump to Paraguarí. I had it rethreaded. I hauled the sucker rod (a steel rod that is part of the surface and downhole components of a pumping system) sixty kilometres round-trip by ox-drawn cart and bus just to have it rethreaded, which took one day of travel and a few minutes of work. No one in the community would have done it, so I did and I succeeded in preparing the pump for installation at the school.

I dragged the sucker rod onto the Medina property at dusk and found most of the family there. Guido arrived with Estela and her husband, Abraham, and their two children. Francisco and Willy came from Paraguarí. Joel and Rodolfo lived there. Some friends of the family arrived. *Ña* Buena found places for everyone to sleep. Privacy was not a consideration. All would stay until New Year's Day, at least.

Sunday, New Year's Eve, I jumped from my bed when I heard agonizing squeals. I ran outside in time to see Noemi's brothers struggling to hold a pig while *Don* Icho slit the pig's scrotum open and extracted its testicles and slashed them free. He held them up in his open palm when he saw me. *Ña* Buena took the testicles from his hand and walked toward the cooking area.

Don Icho changed position and used the same knife to slash the pig's throat. It gurgled and coughed its last breaths as the boys held the pig and one positioned a tub to catch the blood flowing from its neck. With the pig limp, the boys hoisted and hung it from hook on a beam above the tub of blood. *Don* Icho butchered it and passed the skin, cuts of meat, and innards to his sons.

Ña Buena and the other women made sausage from the intestines and blood and other parts. The men hauled off the ribs to a pit to cook and neighbours stopped by to buy the pork we didn't need or want. All parts were used, including the skin, which was deep fried. I ate the cooked pork ribs, but didn't touch the innards or sausage. I avoided eating the testicles, even though the men hounded me to eat. They said eating the testicles would help me keep it up all night.

The next day, we had every part of the pig cooked. From the hooves to the nose and everything inside, *Ña* Buena and the other women made something. They made beans with feet and hocks, soup with innards, roast in the beehive oven, and they deep fried the skin, called *chicharrón*. The men barbequed the ribs and other slabs of meat over coals, and they passed glasses of beer and a bottle of *caña* between them.

The women prepared the tables. They covered them with table cloths and placed covered platters of pork or pots of soup or bowls of pork rinds, salad, *mandioca*, *chipa*, and *sopa paraguaya* on the tables. They piled plates and bowls on each table with knives, forks and spoons. Flies swarmed the table. We served ourselves. We didn't sit at the table, but each filled a plate or bowl and retired to a spot to eat.

I didn't taste everything, but I did partake in the main meal of roast pork, *mandioca*, and *ensalada rusa*, a salad of boiled potatoes, beets, onions, celery, and some mayonnaise. I raised food to my mouth with one hand and swatted at flies with the other. I avoided the soup and the pork rinds.

One table held drinks, including a bottle of *caña*, a few litres of wine, litres of a variety of sodas and some litres of beer. Covered pitchers held wine mixed with orange soda and beer mixed with cola. I shunned the mixtures and drank the Chilean wine. Many in the family preferred the mixtures because of the sweetness.

We spent the day eating, drinking, and sleeping, when one could doze in the shade of the *paraisos* in the oppressive heat of the New Year, a new decade. The previous decade saw the Challenger, live on television, explode and take the lives of seven astronauts. The Soviet war in Afghanistan ended. Military dictatorships ended in Argentina, Uruguay, Brazil and Chile. The Exxon Valdez Oil Spill occurred. The Berlin Wall fell in 1989. Stroessner lived with impunity and in exile in Brazil. The Tiananmen Square protests passed. The Salvadoran Civil War continued. The United States began military action in Panama to oust its president, Manuel Noriega, among many other notable occurrences of the 1980s. We celebrated without a thought for the world outside of Guazú Cuá.

***

Don Icho and Rololfo hitched four oxen to a cart in the fog of the cool, early morning of Wednesday, January 3, 1990.

Ña Buena and *Don* Icho prepared for a trip to Paraguarí to begin construction on a house, and left Rodolfo and Joel with us. The couple had purchased a lot in Paraguarí with the intention of leaving Guazú Cuá, and leaving Rodolfo to care for the livestock. Paraguarí would be a retirement move—as if retirement existed—closer to transportation, electricity, the market, and health care. *Don* Icho would collect his pension as a Chaco War veteran, raise a pig or two, keep several chickens, and would offer accommodations to farmers from Guazú Cuá who transported their products by ox-drawn carts to the market in Paraguarí.

Ox-cart convoys from Guazú Cuá and its environs, upon reaching Paraguarí, had to leave people behind to guard the loaded carts and the oxen. *Don* Icho envisioned an oasis at the end of the trail where owners could enter Paraguarí without worrying about their loaded carts and their oxen.

Don Icho and *Ña* Buena's lot sat just outside of the centre of Paraguarí, near Route 1, and just past the end of the ox-cart route from Guazú Cuá. A traveller could leave a cart and oxen in safety and walk to the market to conduct business. At night the same traveller could store a newly loaded cart in safety and sleep on a cot beneath a mosquito net, inside or outside depending upon the temperature. Before dawn, while men hitched oxen to yokes, *Ña* Buena already would have water boiling for *mate* and something on the fire for breakfast. That was their idea, and they had a great location. It would be a bed and breakfast camp-style.

Progress on the new location would take time. There remained a well to dig, a fence to string, an outhouse to build, and a one-room, wood slab and straw-roofed structure to build in which to live. Each step took longer than one would expect. Materials were scarce and labour unreliable. But *Don* Icho and *Ña* Buena established a home on their property in Paraguarí.

I believed that we would have more privacy at the house in Guazú Cuá, and we did, but *Ña* Buena left her chores to Noemi, which meant Noemi was often occupied. Cows had to be milked every morning, a fire built, the milk boiled, chickens fed, and Rodolfo became dependent upon Noemi for food and clean clothes. I wasn't much help, since I didn't know the animals or how to work them, and the animals knew that. I had plenty to do with my projects, anyway.

***

Storms hit all around on Friday, January 5, 1990. The sky blackened and the wind blew and the air filled with the smell of rain on dust. Heavy rains fell to all sides, but not one touched Guazú Cuá. The *mixto* made it in at night, but not out again in the morning. *Ña* Buena should have come on the *mixto*, but since the *mixto* sat stuck in Guazú Cuá, Rololfo and a friend set out on horseback to the main road to see if *Ña* Buena waited. They rode back to the house on the muddy trail lit by occasional lightning and arrived wet, dirty and hungry after 11:00 pm, without *Ña* Buena. She'd had the sense to stay put.

On Saturday, January 6, 1990, the doctor from GTZ should have visited the site, but didn't, probably because of the condition of the road. At this point, I had given up my key to the *Centro de Salud*, as Dr N_____ stood against me because I wasn't doing what he wanted me to do. Vilma, too, remained irate because I didn't do what she wanted me to do for the school, and she sided with Dr N_____. She had the place to dispense gossip. So without the support of the school director and the support of the *Centro de Salud*, I had to begin again.

That same day, I watched the excitement of barefoot children running from house to house in anticipation of the arrival of the *Tres Reyes*.

27

My relationship with Dr _____ and with Vilma bothered me, as did being an object of local gossip. I was a part of the problem in that I had created conflict by not accepting a role in the doctor's project and by associating with people outside of Vilma's circle. I would always be the subject of gossip as long as I remained the first foreigner to live in the area. Regardless, I realized that I could let go of my feelings of ill will and could try to start anew, and what others said or did lost significance.

On Monday, January 8, 1990, Noemi and I left the site for a couple of days and looked at land around Areguá and Luque with a real estate company, *Inmobiliera del Este*. We thought we might need a spot for a later time. If we ever decided to settle in Paraguay, we would want to be close to the city and potential jobs. Noemi bought a lot in Luque, *Kilometro* 18, for 24,000 guaranies a month for ten years—about eight dollars a month at the time—on high ground and with a view. *Kilometro* 18 is just that far along the railroad tracks from the train station in Asunción, or the Plaza Uruguaya.

Coconut trees—Paraguayan coconut trees which produce clusters of coconuts, each coconut a bit smaller than a golf ball—towered over the corner lot on a high point of the subdivision. Streets were but cleared paths of red dirt and cows grazed freely on the undeveloped lots. The subdivision offered no amenities, except for a communal well about 200 feet below the lot on the way to the main road where busses raced for passengers, leaving black plumes of smoke and clouds of red dust. The high ground and the eighteen kilometres from downtown in the capital city sold the 14.5 x 26.5 metre lot. We thought we might build on it one day.

As soon as we returned to Guazú Cuá, in the evenings, by candlelight, I drafted plans for a basic structure for the lot in Luque. I thought we could stay there on the monthly trips to Asunción instead of paying for a hotel room. I discussed the plans with Jorge. He said his brother might help me and we could visit him. He lived near the bridge that crosses the Rio Paraguay, Puente Remanso, a gateway to the Chaco region.

I left Guazú Cuá on foot with Jorge and Julia on a sweltering Sunday in January. We would go to Jorge's brother's house first to ask about the construction. Jorge and Julia would stay there and I'd go into Asunción to check my mail at the Peace Corps office and to run some errands.

We walked for an hour along the rutted and rocky road, as far as Chircal, when we met a man with a battered, wide-brimmed straw hat, driving an ox cart loaded with ripe watermelons. He stopped his team of oxen and we bought two melons from him. He pulled a long knife from his belt and spilt them for us. We ate on the roadside. The vendor produced a *guampa* and *bombilla* and served himself *tereré* from a jug of greenish water packed with roots and leaves. He offered to share with us, but we declined and slurped and spit seeds from our chunks of red, sweet watermelon. The vendor told us about a murder in Chircal the day before.

"A farmer returned home from his field for lunch, which his wife had prepared for him," he said. "After he ate, he stretched out for a *siesta* in a hammock strung between two trees a dozen paces from the house. His wife took her *siesta* inside. She awoke to a commotion and ran outside where a man with a short, bloody machete threatened her, or so I heard it," he continued. "She ran a quarter mile to a neighbour's house. A group of armed men went with the woman to her house and found her husband dead in his hammock. He had eight stab wounds," he said.

The military apprehended the alleged killer and held him. The victim's sons were in the military, and, according to the watermelon vendor, the military would ask the police for the criminal to be punished without a trial and without publicity. No one knew why the man committed the murder, though many speculated that it had something to do with a woman or a bad business deal.

We thanked the vendor and he cracked his whip and yelled at his oxen and they lurched forward pulling their load. We walked on to the main road and waited for a bus with several other travellers. A blue and yellow bus arrived and skidded to a halt just past us. We ran and squeezed into the bus. I pressed between a toothless woman with a bandana on her head and young barefoot girl who held between her legs a burlap bag containing live chickens. The bus reeked of sweat and dust and animal excrement and diesel smoke. The bus continued to stop for passengers and forced us to ride so close to one another we couldn't move. My groin pressed against the young girl's rear, while the toothless woman's breasts pushed under my arms. I couldn't see Jorge or Julia. At Paraguarí the bus left the dirt and bounced along the cobblestones to the market. It stopped and disgorged its mostly human cargo.

We regrouped and walked past the wood and tin stalls in the market, past mounds of fruits and hanging slabs of meat, to the terminal where we boarded a bus for Asunción. This time we had seats. The bus left the terminal and turned onto Route 1 towards Asunción. It stopped for passengers along the way until the bus filled and Jorge and I gave our seats to women with babies.

By late afternoon, we had travelled about 130 kilometres and crossed Asunción and stopped near Puente Remanso on the Rio Paraguay. We climbed a steep, narrow, cobblestone road to Jorge's brother's house. We found him home and he offered us chairs and served us *tereré*. He couldn't help with the building because of another job he had, but he did recommend someone else in the area who Jorge knew from Guazú Cuá. We looked for him, but he wasn't home, so Jorge said he'd speak with him before returning to Guazú Cuá. I took my leave and caught a bus into Asunción to the Peace Corps office where I checked my mail and looked in a room where volunteers read and left books for others to take. I took a paperback and left for downtown to get a room for the night at the *Itapúa Residencial*.

I preferred the *Itapúa Residencial* to other hotels that volunteers frequented, although it cost a bit more. Each room was different and all were clean, some with private bathrooms and some with shared. I took a room with a shared bathroom because of the lower price. The price included a breakfast of coffee or tea, juice, milk, fresh breads and jams, and plenty of fruit.

That evening I ate alone at an Italian restaurant halfway from the *Itapúa Residencial* to the Baviera on *Calle Estados Unidos*. I stopped at the Baviera for a few beers with some of the volunteers I met there. It seemed as if some volunteers rarely stayed at their sites. Comparatively, I rarely left mine, especially since I'd married Noemi. I'd married into the culture and into the country and acquired brothers and sisters-in-laws, cousins, and an acceptance that I had not known before. I wanted to assimilate, and now many embraced me in that desire. Yet I spoke broken Spanish and I continued to massacre Guaraní.

But sitting at a table in the Baviera with other volunteers, and listening to conversations over beer, it sounded as if I had accomplished nothing while my colleagues spoke of projects and objectives and overcoming difficulties. Some had been in Asunción for weeks waiting for funds or a part for a pump. I drained my mug, paid my tab, and walked to the *Itapúa Residencial* knowing that I would act on my projects upon awakening.

28

As women finished milking their cows, they turned them and the hungry calves loose. The calves scampered about their mothers trying to keep their mouths on teats as the cows gravitated toward grazing grounds. Men shouldered hoes, shovels, machetes, and rakes as they walked toward their fields, their straw hats flopping with each step. Others led teams of oxen to their yokes to pull a cart to a field and then pull a plough through the packed, rich soil. Smoke wafted from each house where girls fried eggs or Paraguayan *tortillas* and heated water for *mate* or coffee. The humid morning air hung heavy with charcoal smoke and cows and manure. Barefooted children on errands queued up at Antonio's to buy or ask for credit for a quarter kilo of sugar or flour or *yerba mate*.

I walked to Antonio's to talk to Vilma. I took a place in line and Antonio motioned to me to have a seat. He gestured that we'd drink *mate*. I nodded and pulled a wooden chair near his fire pit and sat. Vilma came with a bucket of milk to boil on the fire.

"*Buen día*, Vilma," I said.

"*Buen día*," she said, placing the milk on a grill and fanning the coals to life with a fan of woven straw.

"I'm here to see to you, Vilma," I said.

"Why do you want to see me?" she snorted.

"I want to work with you. I want to build the school kitchen," I said.

"The school kitchen?" she said, fanning the coals.

"The school kitchen, yes. If we have one, and we have a protected well and a school garden, UNICEF will donate food for the students every month," I said.

"And you want the school to pay for this?" she said.

"I want the school, and you, Vilma—I can't do this without you—to work with me. I can get a grant—the money—to build the kitchen. But I need you," I said. The original school plans called for a kitchen, but the money ran out before it could be constructed.

"You haven't needed me yet," she said.

"Vilma, I'm regret that we haven't been friends," I said. "But I have something to give and I need your help."

"What sort of help?"

"I need your signature so I can get the grant."

"And after that?"

"Nothing. Not if you don't want to do more. I need you so I can get the funding for the construction."

"If it's for the good of the community, of course I'll give you my signature," she said.

"*Gracias*, Vilma," I said, as I stood to leave.

"Sit, Marcos. Let's drink *mate*."

I sat and drank *mate* with Vilma and Antonio. The past seemed to have dropped off and it was as if I'd entered Guazú Cuá for the first time again. Antonio left his post when he finished the morning rush. Vilma took the pot of boiling milk inside and returned. We passed a *guampa* of *mate* between us and talked of the school, of GTZ, and our break over the films and the handling of the money. Vilma didn't want to handle money. I didn't either. I would have to handle the funds from the grant, but I didn't want cash in my hands from fund raisers. We planned a third party of the other two teachers to manage any money raised for the school.

I left Antonio and Vilma's house with a renewed friendship and an agreement to get Vilma's signature for documents I needed to secure a grant. I walked home and worked on a proposal. I drew a building plan in two views, I wrote a proposal and typed it on my portable, manual typewriter, and I included my community census information. When I finished some days later, I returned to the *almacen* so Vilma could sign the proposal.

***

I found Jorge toiling with a hoe in a field and told him the news.

"Are you sure you can work with me on this, Jorge?"

"I will as long as *Don* Juan works on it, too."

"He said he would. I'll talk to him again. I already included you both in the proposal as builders with your daily pay."

"I'm ready, Marcos."

"It will take some months to get the funding, purchase the materials and get them here from Paraguarí. I used your figures to calculate the cost of materials and added some so we don't come up short."

I left Jorge and walked a few kilometres toward Cerro Verde where I found *Don* Juan ploughing a field behind his two oxen. He stopped at the end of a pass and we sat in the shade.

"I need to know that you will help with the school kitchen, *Don* Juan. I'm sure I can get the funding."

"I told you I'd help you, *Don* Marcos."

"And the daily pay you mentioned is sufficient?"

"I said 10,000 guaranies a day."

"That's right. That's what I planned for in my proposal.

Good. I'm going to Asunción early tomorrow to get this proposal off. We can work on the school well and pump in the meanwhile."

"Whenever you want to start, I'm ready."

I skirted the edge of the field on my way home while *Don* Juan talked to his oxen as they pulled a blade through the topsoil in the field that had been forest a few years before. *Don* Juan felled the trees, had his oxen drag logs to a mill to be cut into boards and slabs, and he piled the slash on the land and burnt it off so he could work the rich soil of the forest floor. *Don* Juan was one of the few who didn't depend upon cotton, although he did cultivate it. He also grew peanuts, beans, corn, and vegetables for his family's consumption. Many locals depended entirely upon their cotton crop for their annual income.

When I reached the house, I told Noemi I'd leave for Asunción the next morning to get a proposal out. I packed my papers so they wouldn't get bent, dirty or wet on the trip and threw a change of clothes in my day pack in case I had to spend the night away. When Rodolfo heard my plans, he caught Gulliver for me and kept him in the fenced yard for the night so I could ride out early.

I went outside when the roosters started crowing and found Rodolfo with Gulliver saddled and ready to go. He saved me hours of chasing the horse to catch it and to saddle it. I scrambled atop Gulliver and bid adieu to the family as we trotted out of the yard and began the fifteen kilometre trek to the main road. Gulliver kept a steady pace and only stopped to drink from the streams we crossed. At the main road, I left Gulliver in a pasture at an *almacen* and I stored the tack in shed behind the store. I told the owner I expected to return the same day and went across the red dirt road to wait for a bus where several others already stood.

Within an hour a bus rumbled up and stopped, raising a cloud of dust. We squeezed between other passengers and found spots to stand on the bus as it roared off toward Paraguarí, where I jumped off at the market and walked a few blocks to Route 1. I caught an express bus, the Quindy, to Asunción and then a local bus to the post office downtown.

I asked to send a stack of letters I'd written by certified mail and the cashier sold me a heap of stamps and gestured toward a group of people crowded about a table. I joined them and moistened the stamps on a sponge in a bowl. There were several on the table and people plastered mosaics of colour on envelopes with stamps of various sizes, each depicting a scene, an animal, a hero, or a painting. I finished my mosaics and stood in line to another window to complete the process of sending my envelopes by certified mail. After my letters were in the hands of a postal clerk, I caught a bus to the Peace Corps office to submit my proposal.

I submitted the proposal for $942.70 through the Peace Corps Partnership Program. The grant award would go to the Peace Corps office to be disbursed to me. I proposed June, 1990, as the month to begin construction.

29

I left a layer of my lip on a metal *bombilla* as I sucked hot water and mashed weeds from a bent, blackened aluminium tea pot. I couldn't breathe through my nostrils and I coughed my throat raw. It was February, mid-summer in Guazú Cuá and I endured a cold. The mashed and boiled weeds would clear my sinuses and my congested lungs, so said *Ña* Buena, who was back in Guazú Cuá on one of her many trips. I knew that the locals used herbal remedies for almost every ailment. When someone prepared *tereré*, a handful of roots or leaves or grasses went into the water, usually for a specific purpose: for the kidney, the lungs, the heart, or for the heat. Even if someone went to a doctor for a sickness, that person would probably visit a *curandero*, a healer, and use herbs and prayer along with the doctor's medication. Most didn't bother with doctors because of the distance to a clinic and because of the price.

The hoot of the air horn of the *mixto* from a distant community broke the peace of the morning. It would pass within three kilometres of the house. It had just begun its service for the cooperative in Jarigua-á to and from Paraguarí. It would leave in the morning and return in the late afternoon, depending upon the weather and probable mechanical problems. A *mixto*, in this case, was a truck with its bed sided with wood and covered with canvas. Wooden benches ran along each side of the bed. The *mixto* hauled anyone and almost anything. People boarded with pigs, chickens, baskets of eggs, sacks of grain, sacks of charcoal, luggage, and anything that needed to be taken to Paraguarí. Two men ran the *mixto*. One drove and the other collected fares and helped load and unload people and cargo. By the time the *mixto* reached Paraguarí, human beings, animals, and vegetable matter all shared the same space.

I ran along a trail of slippery dirt and rocks down to the *Arroyo Curuzú-Ruguá*, rock-hopped across the creek, and ran up again to catch the *mixto*. I jumped on as the truck driver shifted gears. I pulled myself into the back of the truck and met the faces of toothless women, immaculate teenage girls, cigar-smoking men with silver teeth and broad-brimmed cloth hats, and a menagerie of animals and vegetables piled on the floor. I stood for the ride to Sapucaí, as the *mixto* dropped into ruts, spun over rocks and slid in grass. No one dared let go of a handhold. Once we tilted left and another time so far right I heard prayers. We chugged up Cerro Verde in first and second gears to the crest and the wood houses with straw roofs by the roadside, the morning smoke curling from cooking areas, and we dropped toward Sapucaí at a speed that threw us together and apart. Women yelled obscenities in Guaraní at the driver and men laughed. At the toe of the *cerro*, the road levelled and improved and we crossed the railroad tracks, careening into the town of Sapucaí, a cloud of red dust following us.

I climbed out of the *mixto* as the curses and laughter continued. I realized my sinuses had cleared and so had most of my congestion. I paid my fare, dusted off my clothes, and walked to the train station and its adjacent shop to ask about my *fogón* parts. The manager and another boss drank *tererè* in the shade in front of the main office.

"*Don* Marcos! Welcome. Have a seat," the manager said

as he abandoned his seat. "Sit. I'll get another," he said, opening the door to the office. He grabbed a chair, closed the door and sat beside me.

"*Tereré*," the other man said, handing me a *guampa*. I drank and gave it back to him.

"I came to see about the *fogones*."

"You're in luck. I have one for you."

"You have one?"

"One, and it's a good one. I'm going to give it to you, or donate it to your project, I should say. I'm from the *campo* myself, I'm Catholic, and I believe in your project. Do you want to take it now?"

"I'd like to take it today when the *mixto* comes by this afternoon."

"No problem. I'll have someone move it to the stop for you."

"Great. I appreciate that. Where is it?"

"Come on." I followed him into a storage room and he showed me the same sample he'd shown me some months before.

"This one?" I said.

"It's a great one. Well-made."

"What about the others?"

"Others?"

"I need many more. I have people waiting for them."

"I'm sorry, *Don* Marcos, but we are not in the *fogón* business here. Take this as a gift. We wish you the best on your project. If there's anything else we can do for you, just stop by. You know where we are."

"Thank you very much," I said, shaking his hand and his friend's hand. I'll come back later to get this hauled to the *mixto* stop." I walked toward the centre of Sapucaì, my ears burning.

I trudged through town and to the Sapucaí office of *Servicio de Extension Agricola y Ganadera* (SEAG). SEAG was basically an extension agency that worked with farmers and ranchers in an effort to improve methods and production. One woman ran the satellite office in Sapucaí. I'd met her before through Rosa in Paraguarì. She invited me in to sit and chat and I told her about my *fogón* fiasco.

"Marquito, why didn't you tell me? I can get you all the *fogón* parts you want—at cost."

"Seriously?"

"It's not a problem, Marquito. Look," she said, reaching into her desk. "Here is the solicitude. You get the names and identification numbers of everyone who wants a *fogón*, give me the solicitude, and in a couple of weeks I have them all for you. Why didn't you tell me instead of working with those clowns at the train station?"

"I had no idea," I said as I slid the solicitude into my pack. "I really appreciate the help. Would you like to have lunch with me?"

"I would, Marquito, but my boyfriend will be coming for me soon. Just bring me the solicitude as soon as possible."

"I will. I'll have the signatures in a day. We've been waiting a long time."

"Until then, Marquito. I'll expect you."

"It will be soon," I said, shaking her hand as she lent forward so I could kiss her on both cheeks.

I went to a house that served food at lunch and ordered whatever they had that day. It was a steak with fried eggs on top and *mandioca* on the side with a sliced tomato. I drank a bottle of *Pilsen Dorada*, an excellent Paraguayan beer. I stayed at my table after I'd finished eating to write some notes.

I left the dining area and went to the train station again.

No one was there to help me haul the heavy iron to the *mixto* stop, so I walked back to town and hired a driver with an ox cart to help me. He drove his cart to the train station and we slid the stove top and oven into his cart. He dropped me and my cargo off where the *mixto* would soon pass, and it did, within two hours. We loaded the material onto the already full truck and rolled toward Cerro Verde. We crossed the Cerro and at my stop we dropped the stove top and oven at the house so I could return for it the next day.

I hiked down the draw, rock-hopped across the *Arroyo Curuzú-Ruguá*, and walked home. Along the way I decided to build the *fogón* at Nidia's house. She'd been waiting a long time and she was the teacher who helped me the most. I thought she'd probably help me get the names for the solicitude for the at-cost *fogones*.

Rodolfo retrieved the materials for me in his ox cart and he delivered them to Nidia's house. I met Jorge there and told the story to the laughter of all present.

"You can't believe what people tell you, Marcos," Nidia laughed.

"No one?"

"If someone promises you something and you don't see it within a few days, you probably won't see it. Don't pay in advance."

"At least we have one."

"We could have mined and smelted our own ore in the time it took you to get this."

"Well, now I have a better connection at SEAG. We can get all the *fogones* we need, at cost."

"At cost? How much is at cost?"

"I don't know, but less than retail. All we have to do is complete this solicitude and get the names and signatures and identification numbers of everyone who wants one."

"Marcos," Nidia said seriously, "no one will sign the solicitude."

"Why?"

"No one will sign a document for the government, especially for something without a price. That's it. I will ask around, but I doubt we will get a signature. I wouldn't even sign. Would you, Jorge?"

"Not me," he laughed, "I'm already contraband in this country."

"We can try, Marcos, but I doubt anyone will sign, and if one won't sign, the others won't sign on principle."

"If I get the prices first, will that change things?"

"It will help, but it's still a government agency. We have been lied to before. Look around, Marcos. Look at the unfinished school," she gestured, "and the unfinished road, and the unfinished bridges on the unfinished road."

I returned to Sapucaí the following day and went straight to SEAG.

"I need almost thirty *fogones*, but the people will not sign a document before knowing the price and the terms."

"I don't know the price, Marquito. I have to send the solicitude to the central office in Asunción. When the solicitude is approved—and it will be—the *fogones* will be delivered along with the prices."

"I can go to the central office in Asunción to ask the price."

"I'm afraid not, Marquito. We have a strict system, a chain-of-command. You'd waste your bus fare, that's all."

While I was in Sapucaí I squeezed onto a bus and rode to Capiata, a city near Asunción. I went to several hardware stores and inquired about *fogones*. I found them at a higher price than I expected to pay, but they were available and fairly easy to get. I bought a stove top and an oven and carted them home.

30

In March torrential rain fell, a glorious, unrelenting downpour from a blackened sky with water cascading off the straw roof, splashing on the ground and rushing for the already swollen waterways, cutting ruts in fields and roads along the way. I stood in the doorway beside Fatima, who had returned to spend a school year with us. Summer ended. Farmers had harvested their cotton so their barren fields became deep muck. Just a few days earlier, when most were out picking the last of the cotton, I'd gone to an *almacen* near the house to buy flour. It was more of a place to drink, but the woman who ran it, Catalina, also sold flour, canned goods, and other basics. Catalina had several sons and daughters, with a son and three daughters living at home.

"How could she do this to us now?" Catalina said, her silver hair braided and slung over her shoulder. She squatted on a low stool beside her morning cooking fire, beneath the smoke-blackened straw roof of her kitchen. Smoke flowed from the room through the open, slat door.

"So stupid. I shouldn't have said anything to her but I thought she'd understand," she said, wiping her dark, wrinkled nose on a rag she held in her hand, her eyes wide and red and staring at the flames. "It's a hard time for her to be leaving like this with the cotton harvest, and who's going to wash the clothes now?"

"What time did they leave?" asked a young woman, Nelli, a neighbour, who was sitting on an upended fruit crate, cleaning her fingernails with a knife.

"Before dawn, during the night," Catalina said. "It's a good three-hour walk to Acahay. I pray to the *Virgin de Caacupé* they find her there," she said, making the sign of the cross and kissing her thumb.

Nelli stood and placed the knife on a shelf. She grabbed a broom and began to sweep the uneven, dirt kitchen floor. Outside hands clapped. "*Buen día.*"

"*Buen día*, come in," Catalina said, lifting her plump body from the stool and walked from the kitchen.

A dark, smiling, black-haired woman stood in the yard with an empty plastic bag in her hand, as a dozen roaming chickens pecked at the ground near her. It was the volunteer from the *Centro de Salud*.

"Sit down, Lucila," Catalina said, gesturing toward a chair.

"Thanks, but I have to cook. I just came for a half-kilo of flour."

Catalina walked to the small room that served as her *almacen*. Lucila followed her. Catalina opened a wood shutter to allow light to enter. She took a bent and rusty tray from a balance on a shelf and scooped flour into it with a tin can from a fifty-kilo sack. She placed the tray on the balance and added a bit of flour from the can. She dumped the contents of the tray into Lucila's bag.

"Three hundred guaranies," she said.

Lucila handed her a 500 guaraní note. Catalina gave her change and Lucila turned on her bare feet to go.

"I hardly slept last night worrying about Elsa. She left, you know, last night."

"A few days ago one of her brothers, Agustin, came home from the *estancia* where he works. He works with Elsa's boyfriend. He came to tell his father he thought it best if Elsa stayed away from the boy. Says he's lazy and a drunkard and probably won't have a job much longer. Well, that's all her father had to hear. He didn't even know she had a boyfriend. He said he'd beat her good.

"I decided to warn Elsa. Now she's gone. Didn't even take anything with her, no money, no clothes, no shoes—nothing. Now her father's sent her brother and sister out looking for her. Thinks she probably went to Acahay, where the boy's family lives."

I listened and watched as Catalina acted as if I wasn't even there. I interrupted her. "Catalina? Excuse me. I need some flour, too."

"Oh, I'm sorry, Marcos. I thought you came to visit," she said as she dipped into her flour sack. "How much?"

Some days passed and the rain came. The family still hadn't located Elsa, although her clothes disappeared one night, so one of the daughter's, Francisca, got a beating. No one else was suspect. She admitted to nothing. Rumours came and went about Elsa. She made it to Santani or Caaguazu with her boyfriend. Elsa didn't return.

***

As I watched the rain with Fatima, I listened to my shortwave radio. Vice President Dan Quayle and his spouse, Marilyn, visited Paraguay and received a gift of a pair of jaguar cubs named Chaco and Paloma. Animal rights activists protested on the basis of a United Nations resolution condemning the use of endangered species as diplomatic gifts. The State Department itself regulates the importation of such animals. The Vice President and his family asked that the cubs be released into the wilds of Paraguay, from whence they came.

Noemi joined us in the doorway and said that the rain

wouldn't stop until the wind blew from the south, and when it did, thousands of flying ants would come with it. Even if the rain stopped, days would pass before movement began again. Torrential rain and sleep went well together. We all retired for the afternoon.

The south wind did eventually blow and a cloud of flying insects appeared and vanished. The temperature dropped. In a few days, we would host a *novena*, a rosary service that lasts nine days, for Noemi's late uncle Andres, *Don* Icho's older brother. Andres lived ninety-four years. We expected many to attend the *novena*, and many would travel long distances on foot, horseback, or on busses and on foot. Andres had been dead six months.

Customs vary from family to family, but a *novena* occurs at death, again every three months for a year. The second year has two *novenas* and then once a year. This depends upon the family and the economic situation. Some may forego the *novenas* and pay a priest to say masses for the deceased. Others may forget the deceased altogether after the first *novena*.

The first *novena*, at the time of his death, and the second, were held at his home. Now it was our turn, *Don* Icho's turn to host the *novena*. He had to be prepared for a crowd, regardless of the weather. We had no form of communication with the prospective family and friends who might attend. The only form, really, would be to travel to Paraguarí to the radio station and pay to have an announcement made every few hours with the hope that family members would hear and would tell one another. The *novena* stood as planned. Each afternoon of the nine days the mourners gathered and prayed a rosary.

In the Sunday morning chill on March 25, 1990, with the wind blowing from the south, with mud and puddles on the ground, *Don* Icho saddled his horse, mounted and set off. Some hours later he returned leading a young Brahma bull, which would feed the mourners at the *novena*. He dismounted from his horse and tied the bull to a post in the yard. A few minutes later the *Alcalde* appeared and checked the brand and issued a permit to slaughter the animal. *Don* Icho slipped some currency to the *Alcalde* and insisted that he return later to eat. *Don* Icho went for a knife. Antonio arrived with knives, a saw and an axe.

Ña Buena, Noemi, and Fatima constructed an altar of empty soda crates covered with a cloth. On it they placed a photograph of the deceased, Andres, and ceramic statues of saints and candle holders. They finished it with leaves, branches, and wildflowers. On each side they hung blankets to protect the altar from the wind, even though it was beneath the roof of the *galpón*. In front of the altar, they placed a small, wooden table and rows of chairs. The custom called for feeding the participants of the novena. This included a meal and sweets with coffee. *Caña* usually made the rounds, too. The women ground corn and made *chipa* and *sopa paraguaya* and they made cake.

Outside, Antonio and Rodolfo had the bull on the ground with its legs tied and its head tied to the post. *Don* Icho held a pan in one hand and a long knife in the other. Rodolfo kneeled on the bull's back and grabbed its horns, pulling up while the animal bellowed. *Don* Icho shoved his pan under the bull's neck as Antonio slashed the neck wide open. Blood spurted into the pan with each of the young bull's gurgles until life left its terrified eyes. *Don* Icho carried the pan to the cooking area to the women where they would make blood sausage. Rodolfo and Antonio rolled the bull onto its back and Antonio slit it open from the neck to the tail. He reached inside and pulled the warm, stinking entrails from the carcass and tossed them into a plastic tub. The bull yielded 111 kilos of meat, plus the innards.

Rololfo began cutting meat in thin strips to dry. With no refrigeration, the meat wouldn't keep. He hung the strips from a wire beneath the *galpón*. The buzz of flies on freshly hung meat, the stench of the carcass, and the buzzards circling overhead made a scene. Antonio and another neighbour worked the bull hide and stretched it for later use as leather. On rainy days, men like *Don* Icho and Antonio would slice a hide into long strips and then braid strips into rope or they would string the strips in a tight, diamond pattern on a wood frame for a bed.

Ña Buena called us to the table to eat. We sat and the women served plates of bull innards I couldn't identify, except that they were parts I had not eaten previously, or, if I had, it was probably in sausage and I hadn't known the difference. I ate what I could, but I unobtrusively tossed most of my meal to the dogs, aware that I remained an ignorant American. I drank a glass of water, wiped my mouth on the edge of the tablecloth, and excused myself while praising the cooks for the superb meal.

Inside I tuned in my radio and heard that the former Exxon Valdez Captain, Joseph Hazelwood, was ordered to help clean up Prince William Sound and pay $50,000 in restitution for the 1989 oil spill, and that there appeared to be a truce in Nicaragua's civil war. The Hubble Space telescope would be placed into space, something many in Guazú Cuá would not believe, as many did not believe man had been to the moon. I heard the Berlin Wall was coming down. I thought about the distance between my radio broadcast and the people just outside.

I lived in two worlds, one being my life in Guazú Cuá.

The second involved my ingrained beliefs that developed in New York and California and that hadn't prepared me to accept completely the culture in which I had been living in Paraguay. I slipped into my own culture when I read magazines from home or when I listened to broadcasts in English on my short wave radio. Customs such as *novenas* still seemed foreign to me.

Tomorrow the *novena* would begin, and we would say the rosary nine times.

***

The next day, friends of *Don* Icho arrived from Guazú Cuá and nearby communities. The influx of relatives from other areas did not happen because of the condition of many of the roads. Notwithstanding the absence of immediate relatives, we held the *novena*. We prayed, ate, and drank. We ate plenty of beef with *mandioca*. Neighbours brought cake and *chipa* and other food. Slabs of beef from the young bull hung from stakes over a fire pit with several men turning the stakes and sharing glasses of beer or *caña*.

Since not many people attended the *novena*, plenty of meat was left over. *Don* Icho had permission to slaughter the animal but not to sell the meat. He gave most of the meat to people who helped and to people who attended. He still had plenty of dried meat for the family.

I broke a tooth on a bone during the *novena* and, despite the weather, went to the Peace Corps office for help. The nurse sent me to an Italian dentist who performed a root canal and later fitted my fractured tooth with a crown. I learnt during that trip to Asunción that one of my best friends in our group, Steve, had left just a few days before. I met many members of my group while I was in Asunción and I wondered why they were there and not at their sites. It didn't matter. I returned to my site and picked up a stove top and an oven for a *fogón* on the way. Now I had two, plus Nidia's.

31

It had been a year since I first walked to Guazú Cuá and experienced the Feast Day of the *Virgin de Fatima* on May 13, the day of the annual bull fight, one of the top events of the year. Fatima had not seen a bull fight and wanted to go, since most of community would be there.

We prepared Augusto and Fatima and set off for the bamboo and post corral, with a solid chute for the bulls and rickety platforms of bamboo and bamboo posts tied with vines. The platforms bent to the weight of the people on them, those trying to get a good view of the corral. We chose a spot on the ground near the chute.

Men ran the bulls into the chute and cut off the tips of their horns. They let one bull at a time enter the corral where a prospective fighter would harass the bull until it chased him around the corral. After four bulls the handlers, all of whom had been drinking *caña*, fixed on fighting the bulls themselves and wouldn't allow the actual fighters enter the corral. The four drunken, barefooted men swaggered into the corral with a grey Brahma bull, its bloodied and horned head lowered, snorting and kicking up dust. This sufficed for us, so we left. Fatima had seen a bit of the annual festivities.

Any man who aspired could ride a bull before the fighters got to it. For a young man, often in his teens, riding a bull demonstrated virility at substantial risk. I'd watched young, often drunk, riders thrown within seconds and seen more bulls prove their virility than the men who tried to ride them.

We later heard the four handlers were catapulted out and replaced by the fighters. One volunteer mounted a Brahma bull and kicked it with his spurs and yelled at it as it stood passively. A passing heifer wouldn't have aroused the Brahma. The rider slid off the bull to the mirth and mocks of the spectators. He pulled his hat brim low and he disappeared into the crowd.

***

When we reached our house, Rodolfo told me that I'd promised to photograph a wedding in Sapucaí that afternoon and that a truck with the wedding party would soon be by for me. The couple wanted to be married in the church and not the *capilla*, even though a priest had been there that day.

I'd forgotten. Jorge had been the local photographer since he got his camera. He'd left town for a few days and couldn't cover the wedding and I'd offered to do it for him. I knew a few members of the bride's family on a casual basis, but I didn't know the groom at all.

I hauled a few buckets of water from the well and poured them into a five-gallon bucket. I put the bucket in the *baÑadera* and grabbed a towel and a change of clothes from my room. I took a quick bucket bath, dressed, and sat ready with my camera and film. Noemi and Fatima opted to stay home. We heard a truck grinding from first to second gear and back to first. My ride.

A flatbed truck with wood siding and benches in the bed, and with an aluminium frame covered with a green tarp, stuttered to a stop when the driver released the clutch.

People stood on the back bumper of the truck packed with the wedding party and with others catching a ride out of Guazú Cuá. I kissed Noemi, hugged Augusto and Fatima and shook hands with Rodolfo before I walked to the truck, wishing Jorge would suddenly appear. A mother pushed her sons off the bench to make room for me, the photographer, to sit. The driver cranked the truck and stalled and ground gears until he found first and we lurched toward Sapucaí. I scanned the faces of the people packed into the bed of the truck. The bride and groom, apparently, rode in the cab. Men nodded and tipped their hats to acknowledge my presence, and the women greeted me with smiles of gold, silver, and missing teeth, each one with her hair fixed—which only happened a few times a year on such occasions. All wore their best clothing, yet most went barefoot and had their shoes ready to put on when the time came. *Guampas* of *tereré* passed from hand to hand as well as bottles of *caña*. The truck bounced through ox cart tracks, over rocks, and across creeks, the driver grinding gears all the way. We sputtered into Sapucaí and parked near the church. I didn't know it, but we arrived a few minutes early.

Someone tried the church door and found it locked. Others asked around for the priest, but no one knew of his whereabouts. The parents complained about the punctuality of the priest. Some of the men bought more *caña*. Women brushed and braided one another's hair. Children chased each other about and climbed trees while their mothers yelled at them about keeping their clothes clean.

Two hours later a grey-haired, balding priest in a white robe stood in front of the church with his arms crossed. He acted as if we were interrupting his drinking time. He spoke to the families and asked for documentation.

"I cannot marry these two young people. The prospective groom doesn't have his confirmation certificate."

"But you confirmed him," the groom's mother said.

The bride burst into tears. Her bridesmaids surrounded her and wept as one. And one by one each woman began to bawl. The old priest waved his arms and called for an orderly procession into the house of God, God bless him.

The couple married, I photographed the event, and the parents paid the priest. We clambered into the bed of the truck and rumbled back to Guazú Cuá to the reception, arriving eight hours after we left—about a thirty-six kilometre round-trip. I'd had enough of sacraments for the day and opted to skip the reception. Noemi took the camera and recorded the reception and I stayed with Augusto.

32

While I waited for funds for the school kitchen project, I decided to test the 108 children in the local school for intestinal parasites. We had received hands-on training with CHP in Areguá. From the Peace Corps office in Asunción I acquired a microscope, slides, gloves, and other items to conduct a field analysis for parasites, as well as medication to treat those afflicted with the worms. What I didn't have were sample vials, with the exception of several film containers.

I spoke to Vilma, Nidia and Andresa about the project and reasoned that if I found a high percentage of children with intestinal parasites, the parents might be motivated to protect water sources and to build sanitary latrines. They were enthusiastic and arranged for a meeting with the parents during which they explained the project in Guaraní as I stood by with my microscope on a table. I understood little of what the teachers said, but I caught the gist of it. After the meeting, parents swarmed about me for a chance to look into the microscope. I had some slides prepared with dirty water for them to see and I explained that I could identify the microscopic parasites in question from faeces samples from their children.

The following morning I had my laboratory ready before the students arrived. I found out that the teachers told the parents to collect a faeces sample from their children and to have each child bring their sample in whatever container they had available. When a student arrived and gave me a sample, I would label it with the name, and, in a notebook, would record the same, plus the time, the temperature, the type of container, and the relative size of the sample. The school day was split into two sessions, morning and afternoon. Half the students attended in the morning and the rest in the afternoon. About half of the students in the morning session had their samples.

When classes started, I had a collection of almost thirty samples in tuna cans, beer cans, glass jars, plastic bags, liquor bottles, and plastic pill containers. Some samples were a tablespoon or so and others a full load. I ran from the table several times, nauseated, and vomited. I tied a bandana around my head to cover my nose and mouth. I donned my gloves and went to work before the samples dried. I scooped a representative sample from a tuna can and mixed it into a solution, from which I took a drop for a slide. I studied it under magnification. I saw the parasite. I changed the slide for a slide with an example of the parasite and then changed back to make sure I correctly identified the parasite, the hookworm, or its eggs. I counted what I saw and recorded the results. I dumped the waste into a bucket and cleaned my slide and other materials over the bucket. I then picked up a whiskey bottle and fished out a representative sample. The students didn't bring their samples on the same day, so it took four days to prepare and examine 108 samples. Each day I emptied the waste into the school latrine and washed my buckets and other materials in the latrine with water I hauled from a well and heated over a charcoal fire.

Of the 108 samples, ninety-four (eighty-seven percent) tested positive for hookworms. The worms, or the larvae, generally enter through the feet, and most of the children in the area went barefoot. Once in the foot, the larvae enter the bloodstream and find their way to the lungs, where the host eventually coughs and swallows, allowing them to thrive in the small intestine. Eggs leave the small intestine with faeces. Flies, too, have a place in spreading the larvae. Flies swarm in an unsanitary latrine and land on faeces. Larvae may stick to feet. The flies then land on food, leaving the larvae. A person eats the food and the larvae grow in the person's small intestine.

Hookworm infection causes anaemia and protein loss. The loss of iron and protein in children that are heavily infected can result in retarded growth and mental development.

I gave a report on the results to the teachers. They, in turn, spoke to the parents and recommended that they get the free medication from me for their children. Some did and some didn't see it as a problem, or believe it was a problem. Others said it didn't make a difference because the children would still go without shoes and flies would always be attracted to faeces and food.

***

While I waited for my grant I spoke to Nidia about fundraising activities we could do for the school. Before Peace Corps I danced a few nights a week with an International folk dance group and with a group of recreational country and western dancers. Nidia danced and she sang, too. So did Noemi. We knew a few guitar players who also sang, including he who would be the best man at my church wedding, *Don* Juan. Jorge did tricks. One was to place his head on the back of one chair and his ankles on the back of another. With his arms at his side as if he were at attention, he remained erect between the chairs with nothing beneath him. I had a book of one-act plays written by Paraguayans. We decided to start a performing arts group.

We recruited eighteen performers and most of them had never sang, danced or acted. We arranged to practice in front of Nidia's house a few nights a week. I used a cassette player powered by my solar powered Peace Corps battery that I had for the slide projector.

Teenage girls fidgeted in chairs while the boys stood away from the house and talked. The older members of the group sat by the cassette player and listened to "Walk of Life," a Dire Straits song I would use to teach a swing dance.

I went from girl to girl to introduce the basic step until I found that Zuni, Nidia's daughter, had natural rhythm. She acted as my partner for all demonstrations. We practiced the basic swing steps to the same song dozens of times. Nidia arranged traditional Paraguayan dances to break from the swing. She choreographed dances to live music and songs for a few couples and a dance for two girls and a solo dance. Juan's daughter, Luz, did the solo dance barefoot and in a long, spacious Ñandutí skirt and a hand-woven cotton blouse called *ao po'i*. The lyrics of the song for the dance, in Guaraní, tell a story of a young girl who has been thrown out of her parents' house for her behaviour.

We practiced when we could and changed houses to make the burden of walking long distances on chilly autumn nights as equal as possible for all involved. Some of the eighteen always missed practice, but not the same ones each time. Juan and Desi always made practice with their guitars and a canteen of *caña* and roots and leaves, which they guarded and sipped at throughout the practice. Juan unobtrusively offered me a drink on occasion. I didn't understand the effect of the herbs in the *caña*, but the mix packed power and apparently Juan's recipe involved soaking the herbs in the *caña* in a dark spot for a few days before the concoction could be used. He kept jars of the liquid in rotation so he always had a ready supply. I never saw the man drunk or even buzzed, despite his hobby. Desi did allow the drink to alter his spirits and often arrived at practice quite animated. Juan and Desi played guitars and sang, usually in Guaraní. Silvio, a local carpenter and guitarist, also joined the group.

Noemi and Nidia choreographed a song and an act. Each verse expressed a point of view from a woman, Noemi, or a man, Nidia. Both joined in the refrain. Nidia dressed as a man, a Paraguayan farmer, and Noemi as a country girl questioning the motives of the man. Juan, Desi and Silvio played their guitars. Along with our swing, Paraguayan folkloric dancing, and the singing and acting, we selected a one-act play for Jorge and Julia about a man who arrives home with lipstick on his collar. After a hot and humorous argument, at what appears to be the climax, the son arrives home and says that he had used his father's shirt the night before. We did a few other one-act plays as well, and, of course, Jorge did his metaphysical magic. We had enough material to entertain a crowd for an evening.

We weren't very good. Most of us couldn't keep a beat, but the audience didn't notice or care. We had the only show in town. We sold food and drinks and charged an admission. I considered it my most successful project, even though I drifted from my primary project of environmental sanitation. The funds we raised went to the school and not to environmental projects. By that time, I had become used to drinking parasite infested water, eating contaminated fruit, swatting at flies with one hand as I ate with the other, and considered a solid bowel movement as a remarkable event. I entered Peace Corps at 150 pounds, probably underweight, and left at 125 pounds.

One night as we gathered for practice, I placed the cassette player, the charged solar battery, and our tapes and written material on a table. One of the girls connected the battery to the cassette player.

"Marcos! The battery!"

A sulphur-like stench permeated the area before I turned to see the battery smoking and deforming. I ripped the crossed cables off the battery. The girl who connected it cried. I assured her I'd replace it and we could practice without the taped music. I did replace the battery on my next trip to the Peace Corps office and I had $108 deducted from my readjustment allowance.

But with the nights longer and often chilly, with the winter break from school at hand, and with an increasing frequency of torrents of rain which stranded people in their homes and close roads, we suspended practice and plans for performances until after winter break.

33

Fatima first arrived in Guazú Cuá in an ox-drawn cart driven by a friend of the Medina family. The cart's dusty wooden wheels ringed with iron spanned a diameter of several feet to manage the often muddy, always rutty roads. Fatima sat in the back of the cart with her dusty brown legs and bare feet dangling over the edge, her brown hair tied back with a strip of cloth. A flour sack beside her contained her clothes. Her brown eyes stared back at the way she'd come.

She came without shoes, little clothing, and no toothbrush. She didn't seem to miss what she lacked, but we wanted more for her than her patched, oversized skirt and her worn, undersized blouse. Fatima often described her home, Durazno, where the trees grew heavy with avocados, mangos, papaya, oranges and bananas. Cotton grew tall on the hilly fields. Children pulled fish from Durazno's streams. Fatima dreamt of going home, even though at home she toiled in the fields, gathered fire wood and cared for her siblings. She wanted to go home for winter break.

We'd never been to Durazno, but we knew it to be about 180 kilometres to the east in a recently settled area away from a main road. We told Fatima we'd take her if she knew the way to her house from the nearest bus stop.

"If we leave early, we'll be at my house for lunch," she smiled.

Winter recess arrived after several days of heavy rains.

The municipalities closed unpaved roads during torrential rains to avoid damage but this made travel difficult. Fatima accepted that the trip may be cancelled. She stood in the doorway staring at the saturated earth.

Creeks swelled to their high banks. Our well filled to ground level. Gushing water wore at the ruts in the mucky road. Frogs chanted. A sheet of water covered the ground moving toward creeks and drainages. Torrents of rain lashed the house, and water found its way in through the roof and walls. One day a cold wind blew from the south, a sign that the rains would end. After a two-day dry spell, we decided to travel.

Before dawn Noemi, Fatima, Augusto, and I plodded through muck and puddles on the way to the bus stop. The mud slowed our walk, as did the swift-running creeks we crossed on makeshift bridges. Fatima sang as she walked and acted as if we were bound for paradise. I carried a backpack with all our gear. Augusto sat atop the pack and held my head. On the road we met a group with four oxen dragging a cart through axle-deep mud. Men on horseback led the oxen, whipping them to pull harder and faster. The cart carried two women. One moaned and the other tried to comfort her as they lurched and slid toward medical help.

We reached the bus stop at the main gravelled road after three hours. We slipped into some brush to change our muddy clothes. Soon the eastbound bus arrived empty. The driver said that he'd be going another forty kilometres and then returning. There were no through busses because high water swept away a bridge along the route. We waited for the bus to return to take it westbound toward Asunción. We could still make the trip by a roundabout route.

We caught the bus to the terminal in Paraguarí, ate some *empanadas*, and then boarded a second bus to the terminal in Asunción. Fatima had never been to the capital, had never seen so many people walking, waiting, working, hawkers on the streets selling watches, fruits, handkerchiefs; barefoot boys yelling: "*Ultima Hora*! ... *Noticias*! ... *ABC Colour*! ... " the names of newspapers; women balanced baskets on their heads yelling, "*Chipa ... Chipa ... Rica Chipa.*"

At the terminal, we bought tickets to General Morinigo, a town south of Villarrica and the closest to Durazno. Our bus left packed floor to ceiling with people and baggage. We got one seat and took turns in it during the trip. The bus stopped at major towns along the way. Vendors swarmed the bus at each stop to hawk their water, yoghurt, fruit, soda, and *chipa*.

The pavement ended at Villarrica. After a stop at the terminal, we continued in the winter darkness sliding and jolting along a rutted road for two hours. Only we four got off the bus at General Morinigo, a town that had been asleep hours before our arrival. We watched the bus rattle off and stood on the main street of August, General Morinigo in total darkness. I lifted the pack onto my back and then lifted Augusto atop the pack.

Fatima bolted like a colt out of a corral. She held a flashlight and led us around puddles and mud along a wide trail slashed through a growth of towering hardwood trees with thick vines hanging from them and intermittent soaring stands of bamboo and wet, low brush. The path ended abruptly several times and Fatima found its new beginning each time after a search. We walked to the drone of insects, the rioting of frogs, and the scurrying of unseen creatures of the night. Bamboo stalks rustled in the cool breeze blowing from the south. Passing clouds shaded much of the moonlight. The dense air smelt of rich, composting earth. Before us spread cultivated fields dotted with coconut trees.

"*Ja guahe po ta ma* (we're almost there)," she said in Guaraní, without breaking the pace. We trudged another two hours in fields, along ridges, through draws and across creeks. Augusto slept atop the pack so I held him with both hands so he wouldn't fall. I tired from carrying the pack and doubted Fatima. The fields and creeks we crossed looked familiar. As we passed a thick, black stand of trees that I was positive I'd seen before, Fatima pointed her dimming flashlight into an orange grove, beyond which stood a hovel on high ground beside a stand of bamboo that broke the cold south winds.

"*Mamá! Papá!*" Fatima called.

"Fatima?" responded a woman's voice. We then heard other voices and saw a small flame flare through the spaces in the wall and we heard a door creak open. Fatima's mother, Cristina, stepped outside holding a lit candle.

"*Pe guahe, pe guahe* (come in, come in)," Cristina said, standing in a corridor between two rooms, both built with coconut bark and bamboo with a straw roof.

"It's Fatima. Fatima!" she said, as she stood barefoot on the red dirt floor. The flickering candlelight illuminated her dark hair hanging on her wrinkled dress. Her green eyes flashed from her bronze face in the candlelight and she wore a smile that was, in itself, worth the trip. We followed Fatima into the corridor.

"Fati, *e jew koape xe memby* (come here my child)," she said.

Fatima pocketed her flashlight and stood before her mother. She clasped her hands together as if in prayer, a sign the young gave in reverence to their elders, such as parents and relatives. Cristina extinguished the candle flame and made the sign of the cross with her fore and middle finger on Fatima's forehead. "*Dios te bendiga* (God bless you)," she said, and then they embraced one another.

Cristina turned to Noemi and they shook hands and kissed one another on both cheeks. She did the same with me as Fatima introduced me, although she already knew me by hearsay.

Six children spilt into the corridor, four from one room and two from another, and smothered Fatima with hugs and kisses and questions. Cristina realized I carried a pack and told me to put it on the table in the corridor. I lowered Augusto, now awake and too confused to respond, and Noemi took him into her arms. I wiggled out of the backpack and let it drop on the table. Energy surged through me as I freed myself from the weight that had felt a part of me during our trek.

"Fati! *e jew koape*," called a man's voice from inside. "I can't get up." Fatima stepped into the dark room to see her father, Enrique.

"He stepped on a broken bottle yesterday and drove a chunk of glass two inches into his foot," Cristina said.

Fatima called to us to enter the room to see Enrique. Cristina said she'd fix something for us to eat. She left the corridor and ducked into a low lean-to behind the house to build a cooking fire. Fatima rejoined her siblings in the corridor and we entered the room to greet Enrique.

An unsteady flame flickered from a candle stub on a bed post. We stood in a space on the dirt floor between a bed and a cot. Clothing and tools hung from the walls and rough-hewn rafters. Hoes, shovels and machetes lent in one corner. On the wall opposite the door a dusty shelf held a statue of a saint encircled by short candles, beads, and bright strips of cloth. Enrique, a dark, lean, bearded man beneath a blanket, reclined on the bed. He tugged at the blanket so we could see his rag-wrapped foot and apologized for not getting up. I shook his hand and Noemi kissed him on both cheeks while she held Augusto. Cristina called us from the kitchen.

We left Enrique and ducked into the smoky lean-to with Cristina, her hair tied back with a strip of leather, squatting beside a fire burning within rocks and covered with blackened metal rods tied together with wire. She fried *mandioca* and offered us peanuts to eat while we waited. We sat on the floor and ate peanuts and told of our trip.

She laughed as she turned the *mandioca*. "If you had stayed on the bus for one more stop your walk would have been much shorter and easier. You would have been here hours ago," she said. Fatima had known she was close to home, but hadn't known exactly where we were until we happened upon her house.

Cristina dumped the sizzling, golden *mandioca* into a large bowl and pulled a sack of hard rolls from a nail on the wall. She placed the food before us and grabbed a pail to fetch water from a nearby stream. We washed our faces and hands with the cold water and then ate. Fatima stayed with her brothers and sisters.

After we ate and talked, we returned to the corridor and Cristina herded six of her offspring into one room. Three crawled into each of the two cots and buried themselves beneath the covers. We followed Cristina and her youngest into the other room. We had the cot to ourselves.

"I'd like to use your outhouse before I get into bed," I said.

"Anywhere away from the house is fine," Enrique laughed. "We haven't built one yet."

I walked into the orange grove and thought about my own alienation from nature as I squatted on the wet ground. I had been accustomed to sanitary bathrooms with flush toilets and privacy. But in Paraguay, I attended to my basic needs in fields and woods where indoor plumbing did not hide the stench of my own excrement, my own nature.

I returned to the house and made myself comfortable beside Noemi. Cristina snapped her fingers on the candle flame. The chilly night air passed through the cracks in the walls. The room reeked of mildewed straw, earth, sweat and cigar smoke. Outside bamboo stalks rattled in the breeze.

"Did you get enough to eat?" Enrique said. "Are you warm enough? If you need anything, let us know. We're glad to have you here. Tomorrow we'll celebrate. We'll kill a chicken."

34

I watched rain fall for most of July as I assessed my Peace Corps experience. I had been living in Guazú Cuá for about fifteen months. My service had begun eighteen months earlier with a swift yet violent change of government, which had changed little if anything for the average person. I'd fallen in love with and married a woman from my site, an act that kindled the flames of gossip within my site and the Peace Corps community.

I questioned my ability and willingness to continue living in an undeveloped area without basic services. I had wanted to assimilate into the community of Guazú Cuá, to meld my culture with Paraguayan culture. My wife could live with or without modern conveniences. I didn't know if I could live without them as if I were on an extended camping trip. I also contemplated my unfinished projects and my ideas for projects, and I considered the definition of success. What did success mean to me?

Success meant doing the best I could for the people of my site. It meant understanding another culture. It meant living without indoor plumbing, electricity, a telephone, and motorized transportation until the end of my service. And it meant maintaining a positive attitude regardless of what the doctor and the school director said or did. Success had nothing to do with the acquisition of anything except acceptance of my immediate state.

Meanwhile, water puddled on the saturated earth.

During breaks in the rain and such thoughts about my experience, I walked out of my site a few times to go to Asunción to check on my school kitchen grant and to look for the parts I needed to make *fogones*.

The approved grant travelled so slowly from the United States I wondered about its reputation as a developed country. Students from a private school raised the money and the United States Agency for International Development (USAID) administered the small grant program.

I also attended a Peace Corps Fourth of July Conference at San Bernardino beside *Lago Ypacaraí*, northwest and across the lake from Areguá. The conference was mandatory, though I believe no protocol for disciplinary action existed if a volunteer failed to attend.

Almost every volunteer in Paraguay and the support staff gathered at a hotel that was inhabited by Paraguay's affluent during the summer. We listened to success stories and to updates about Peace Corps Paraguay from cushioned seats beneath chandeliers. Motivational talks by volunteers concerning their experience found groups slipping out to the bar, pool, or for other entertainment. I enjoyed the talks and the entertainment and the meals at the hotel, yet held back from the excessive drinking.

At the conference, on the eve of the Fourth of July, one volunteer I'd never seen before rushed dripping wet into the lounge with blood gushing from his head. A group followed him, some trying to help him while others told his story. He refused help, plopped into a stuffed chair and held a handkerchief to his head.

"Just get me a beer and a couple of shots," he said. "I'm okay, really. I thought it was the deep end and I hit bottom quickly."

"Is anyone else in the pool?"

"No. It's cold and I just took a dive."

"It is winter, you know?"

"I know," he said taking a beer in one hand and a glass with a few fingers of whiskey in the other. He'd tied his bloody handkerchief to his head. "It was stupid but I wanted the experience."

"He's a doctor," someone said.

"I'm not a doctor, but I finished medical school. I'm doing Peace Corps before my residency. So I know I'm okay," he said, draining his glass of whiskey.

I walked to another lounge where a volunteer tottered on a stool before a small audience. A couple of harmonicas soaked in a glass of vodka on a table beside him. He played blues as he swayed on the stool. I heard some months later that during the conference he was still in training and that he had volunteered with his wife. The family with whom they lived reported him for domestic violence which earned him a ticket home and the end of his marriage.

I'd spent most of my time at my site and had had little contact with other volunteers. I'd been living a very different life than many of my colleagues in that I had married into the culture and, in doing so, had circumvented contact with other Americans. I hadn't used any of my leave time, while many in my group had travelled to Buenos Aires, Santiago de Chile, Tierra del Fuego, and Rio de Janeiro. Travel had been one of my objectives, but it was as of yet unrealized.

After the conference in San Bernardino (which ended with a meal of beef tongue—an unpalatable dish for many of the volunteers), most of the group left for Asunción to continue the party or for other reasons. I checked my mail and the status of my grant at the Peace Corps office and boarded a bus to get back to Guazú Cuá. Some hours later

I arrived in Paraguari, where I'd catch another bus to the road to Guazú Cuá. It hadn't rained for a few days, but the sky blackened with clouds at the onset of night. I sat in a restaurant and ate while I waited for the last bus. I met some others from Guazú Cuá and they said they'd stay with relatives because of the time and the weather. They implored me to stay with them and we'd all travel together the next day. But I had home—Noemi and Augusto—on my mind and refused to listen to reason. I knew I could make the walk in just over two hours. I'd done so before but during daylight on a dry road.

I hadn't achieved the ability to wait as the Paraguayans could, nor had I attained the concept of time natural to the locals. I believed my punctuality a virtue and considered waiting synonymous with squandering time.

I boarded the last bus and stepped off at my stop as dusk fell. I shouldered my pack and set off along the wet, red road as blackness replaced the grey dusk. I pulled a small flashlight from my pack and used it to scan the road ahead. Thunder rolled and lightning illuminated the road before I felt the first drops of rain on my face. My flashlight dimmed to the point of uselessness so I put it back in the pack. The few seconds of the first wave of the storm drenched me. I'd gone too far to turn back and the ox-cart ruts in the road filled with water and the red dirt became as slippery as ice. I pushed on through the torrents of rain as I slipped into one rut after another and thought of animals that might be watching me slosh through their habitat. I only knew I walked on the road because I wasn't in the brush or woods. Booms with bolts of lightning helped me to stay on course until I reached Guazú Cuá and home, wet and covered in red mud from my stomach to my feet. The experience renewed my faith, though I questioned my decision to make the walk. The rain had continued throughout my four-hour walk, double the time I had anticipated.

***

School started at the end of July as the weather changed. Every morning as I thought about projects and plans, I'd watch barefoot students, some with shoes in hand, pass the house across the saturated ground on their way to class. Daytime temperatures increased yet with nights cool enough for pleasant sleeping. Guazú Cuá awoke from its winter slumber and new shoots and leaves gave green life to the *paraiso* trees in the yard and some lapacho trees bloomed brilliant lilac from the fresh luxuriance of the wooded areas. Clusters of mushrooms materialized overnight in Guazú Cuá's common pasture where grey Brahmas, dirty white sheep, and horses grazed by day. Smoke from cooking fires hung in the air and its odour mixed with the smell of cow dung and flowering plants in the otherwise fresh air. Women walked to and from Antonio's store and men rode horses, drove ox-drawn carts, or walked to work, hoes on their shoulders, in fields of cotton or grain, tobacco or peanuts.

I read from a list of projects I planned to accomplish during these last months of my Peace Corps service. I would organize fund-raising activities for the school, construct *lozas* for the floors of sanitary outhouses, and would build the school kitchen when I received funding. I also planned to continue to collect books from embassies for the start of a school library and to continue with the construction of *fogone*s. I'd start a dental health program in schools in nearby communities as well as my own. I aspired to get community support and a grant to fund a well, pump and an elevated tank from which water could be gravity fed to most of the homes in Guazú Cuá.

***

I started with a trip to the Peace Corps Office to inquire about my school kitchen grant and then went to SENASA for information about the *lozas* it had been producing. My projects and plans changed, though. I found, instead, a letter from my sister, Fran, telling me that my father was very sick and, apparently, had colon cancer. My father's second and last wife practiced Christian Science, and he followed her advice in all matters. He had not sought the opinion of a medical doctor until he had been sick for some time. When he did, the doctor diagnosed him with cancer, presumably of the colon.

35

I requested and Peace Corps Paraguay granted me emergency leave and provided me with a round-trip ticket from Paraguay to Sacramento, California. I called home from the Peace Corps office and told Pat, my step-mother, that Peace Corps made arrangements for me to travel and that I would let her know my itinerary. She assured me that my father wasn't too bad and it wouldn't be necessary for me to take emergency leave. I ended the call, but then soon called home again and arranged to have my step-brother, Tom, meet me at the airport.

I returned to Guazú Cuá and told Noemi the news. I'd be gone for at least two weeks. I packed no luggage except a carry-on bag. I walked out of Guazú Cuá the next day as rumours spread about my sudden departure and that I would not be seen again.

Tom met me at the Sacramento airport on a warm evening, the beginning of my emergency leave, and we drove an hour northwest to the house in Grass Valley. We talked about family, my father and his disease, and we listened to talk radio. President George H.W. Bush had nominated Clarence Thomas to a seat on the Supreme Court and his forthcoming hearings kept talk radio busy with an allegation of sexual harassment from Anita Hill.

I hadn't been out of the country but a year and a half, yet I experienced reverse culture shock during that ride, listening to talk radio, listening to the hum of the tyres on the finely surfaced road, and watching the dizzying neon lights as we passed fast-food restaurants and stores selling just about anything people might believe they needed. I thought of Noemi and Augusto and the fifteen kilometre walk to Guazú Cuá on a road suitable for ox-drawn carts, horses or foot travel, and occasionally four-wheel drive vehicles during dry weather.

We turned into the driveway and stopped in front of the house. Pat came out to meet us and my father stayed inside where he sat and read. After greeting Pat, and after explaining that, yes, I really did have only a carry-on bag, I entered the house to greet my father, Marko, after not having seen him for a few years. I'd lived in New Mexico when I joined Peace Corps and I hadn't gone home often during my eight years in Albuquerque and Socorro.

He dropped his book, took off his glasses, stood and we hugged. I felt the loose flesh on the flimsy body of the man who spent most of his life swinging a hammer. I spent the next two weeks with him. I dug for the top of the septic tank, I uncovered pipes, I crawled under the house to insulate pipes, and I drove him to his first couple of chemotherapy treatments. He quit the treatments after the first few. We drove to town and visited his favourite coffee shops and I completed the paperwork for a building permit for a rental cottage Pat wanted to build, and did build, on the property.

My sister Fran arrived from New York for a few days while I was there. We looked at hundreds of photographs and split them between us. Pat didn't want them. I left mine there to pick up after Peace Corps. After Fran left I did more chores around the house.

One day I drove to the county buildings to acquire the building permit. On my way home, a California Highway

Patrol cruiser passed me, made a quick U turn and tailgated me with lights flashing. I pulled over to allow him to pass me, but he too pulled over. He wanted me.

The patrolman walked to my window, one hand resting on the handle of his pistol, and asked for my license. As he looked at it he said, "Where's the camper?"

"Camper? I have no idea."

"The camper. You're supposed to have a camper. Where is it?"

"I'm sorry, Officer, but I don't understand. Why am I supposed to have a camper?"

"Your plates indicate a camper."

"I'm sorry, Officer, but this isn't my truck and I really don't know what you're talking about."

"I'm talking about taxes. You're supposed to have a camper," he said as he wrote a ticket.

When I got back to the house, I told the story. The camper was in the garage and taxes were less if a truck had a camper. My father explained this as he ate ice cream from a gallon container and took bites from a doughnut.

As we talked about the ticket, Pat chased and swiped at a single fly. "Help me get this fly. How did a fly get in here?" she said, swinging her swatter here and there.

"Pat," I said, "where I live I often eat with one hand while I wave flies away with the other. I probably eat several a day. One fly in the house doesn't upset me."

I spent some time with my maternal uncle, John. He bought clothes for me to take to Noemi, Augusto and Fatima. He said he thought the doughnuts and ice cream were keeping my father going, since they helped keep some weight on him.

During my stay, I watched my father try to do things for himself, like use the bathroom, but fall in the process.

I read some of Ring Lardner's short stories to him to try to get him to laugh. I saw him sob in anger because he couldn't do things for himself and because he'd just retired at sixty-six and now faced death armed with nothing but a religion in which he didn't believe and plenty of costly painkillers.

My father told me I should leave, to go back to Paraguay with my family, back to my life, yet he wanted me to stay. He wanted to meet my family before he died. I thought I'd better start working on visas for Noemi and Augusto when I got back to Paraguay. He might just meet them before his death. Before I left my father gave me the wedding band he'd given to my mother and the band that Pat had given to him. He thought he'd never see me again. I wished Noemi and Augusto had been with me to meet him.

I spend my last night in California at Tom's house in Sacramento and he drove me to the airport in the morning. I boarded the plane with my overstuffed carry-on, glad to leave but sad to leave behind my father who I might have seen for my last time. I had plenty of time to think about it as it took a few flights to Miami and then another series of flights from Miami to Panama to Peru to Bolivia and finally to Paraguay.

I spent a night at a hotel in Asunción and went to Guazú Cuá the next day, watching whispering women at each dwelling I passed, their rumours dispelt. I hadn't vanished. As soon as I stepped into an area where I could see the distant house, Noemi and Augusto ran to meet me, her with joy-filled tears flowing and he laughing. I knew two weeks with my father had been sufficient, and that I was home again.

36

Eighteen of us met at Nidia's house three nights a week to practice our show. Our ages ranged from fifteen to fifty and we choreographed performances for fund-raising events. After some weeks of practice and planning, we advertised our first performance at a house with a grassy area in front. We set up a *cantina* to sell soda and beer and snacks. We made a circle of benches with slabs of wood and stumps. We hung some cloth as a backdrop for the play and for Jorge's trick, and to slip behind to change costumes. We had our cassette player next to the cloth and some chairs for the guitarists. Some of us baked cakes or other sweets for a raffle.

Guazú Cuá's Amateur Performing Arts Group joked beside the cloth as spectators drifted in, made purchases from the *cantina*, and found places to sit. Juan, Desi and Silvio tuned their guitars and strummed quietly. Teachers collected an entrance fee and initialled the wrists of those who paid with an indelible marker.

Nidia yelled to the crowd. The drone of voices gradually quieted. "Thank you for your support," she said, smiling beneath a straw hat. "All of our profit goes directly for school improvement. Our group consists of all volunteers. Get your food and drinks at the *cantina* and we'll start in a minute." The audience clapped, most sat but some stood at the *cantina* sharing beer.

The music boomed from the cassette player and out we

danced, a couple at a time until six couples formed a circle within the circle of onlookers. We flowed between country swing and two-step, spinning, twirling, and changing partners on the sloped, uneven grassy stage. The audience roared with applause as we missed steps, danced out of time, and stumbled on the turf. No one had ever seen anything like it and the people loved it. We bowed to our partners and then to the crowd and followed with a fast waltz and then a polka. We danced to rock and country music I selected. As bad as we were, we must have looked good in our hats and traditional Paraguayan garb. The women wore long, flowery dresses and knitted cotton blouses. Men wore wide, long, striped red, white, and blue cotton belts wrapped several times about the waist and tied in front. Most danced barefoot. We bowed to the audience again and ran off the grassy area to a standing ovation.

Juan, Desi and Silvio walked with their guitars to the centre of the circle and sat on chairs that had been placed there by a couple of young helpers. They performed a few folk songs in Guaraní and then Noemi and Nidia joined them. Noemi wore a dress and Nidia wore men's clothing and a moustache. The guitarists played and Noemi and Nidia sang a song in which the man tells the woman how good he is. She disagrees and proceeds to explain how much better she is, and the dialogue, in Guaraní and heavy with sexual overtones, continues until both decide that each is good enough for the other and they agree to spend their lives together.

Jorge and Julia took the stage and did their one-act play and then Jorge did his gravity-defying stunt on the chairs. The audience listened and laughed throughout.

The group did another improvised play. A man enters a house and introduces himself as a veterinarian. The woman of the house doesn't know what a veterinarian is, but she needs a doctor. The vet sees an opportunity to make some money and proceeds to examine the woman. Word spreads that a doctor has arrived in town and people line up outside the woman's house. By and by someone realizes the medication the doctor has been selling to the people is for livestock. This performance ended differently each time, but always to applause.

We ended with a few couples doing Paraguayan polkas and with one of Juan's daughters performing her solo dance. The six couples came back for one more swing and then the group of eighteen gathered on the green stage and bowed. The show was poorly executed but the spectators didn't know the difference. No one had witnessed such a live performance before.

We performed about once a month in Guazú Cuá and in nearby communities, and it was always successful in the eyes of those who watched us.

***

Dr N_____ continued with his GTZ contract. He seldom appeared in Guazú Cuá, but he did, at times, send envoys, such as a nurse. One morning several of us sat in the shade beneath the *paraiso* trees and drank *tereré* with Roldolfo serving. The *guampa* came around every few minutes. We watched livestock graze in the common pasture of the community. Chickens scrambled about pecking at the dirt for insects. A slight breeze kept the country air fresh, except for wafts of manure from the corral or wood smoke from a cooking fire.

"Someone is coming," Noemi said, gesturing toward the edge of the pasture at the wood line about a kilometre distant where the road entered Guazú Cuá. All eyes turned.

"It's a woman."
"She's carrying something."
"Sure walks slowly."

I strained my eyes and still couldn't see anyone, not even a movement in the distance. The others speculated on who she could be, what she carried, and why she would be walking alone to this community. Someone grabbed my head and tried to aim my face toward the figure obvious to everyone but me. I did see her several minutes after the others. I had good vision but I just hadn't sat and observed long enough to detect a slight difference in the landscape, such as a white speck moving slowly. I knew then that the others could spot me the moment I exited the wooded area and they could tell what, if anything, I carried. A person must study their surroundings to be aware of minuscule, distant movement. I lacked the ability to simply sit for hours and to observe my environment, possibly the result of the culture in which I had spent most of my life.

"It's an ice chest."
"She's wearing a nurse's smock."
"Must be going to the health centre."
"Doesn't move too quickly."

She dropped out of sight as she crossed a drainage area and when she emerged she walked as if she came to visit us. She limped right into the yard, half dragging her dusty ice chest, her white smock smudged with signs of her journey, her bangs matted to her forehead with sweat.

"Welcome," we said.

"Thanks," she said, dropping her ice chest and sitting on it. She pulled a pack of cigarettes from the pocket of her smock and waved it before us. "Cigarette?" She said as she pulled one from the pack and lit it.

"No, thanks," we said.

"How far did you walk?"

"From Chircal. I got a ride to there," she said, blowing smoke. "Are you the Peace Corps Volunteer?" she said, gesturing with her cigarette.

"I am," I said. "I'm Mark, or Marcos. And you?"

"Carolina," she said. "Dr N_____ sent me here to vaccinate children."

"And he mentioned me?"

"He said you'd help me."

"I will," I said. "The health centre is over there," gesturing toward the deserted building. The volunteer had stopped going to the *Centro de Salud* and Gaspar only went when the doctor was scheduled to visit, which wasn't often.

"I wouldn't mind having a beer first," she said. "Can I get a beer around here?" she said digging in her pocket for some guaraní notes.

"I'll get you some," Rodolfo said. "What do you want?"

"A litre of *Pilsen*, please," she said, handing him some money. He walked off toward Antonio's store.

37

My grant money for the school kitchen arrived on November 14, 1990, just weeks before the semester ended. I used that time to purchase materials, with the help of Jorge Encina, and arranged to have them delivered to Guazú Cuá. Delivering to Guazú Cuá proved challenging. One truck hauling a load of bricks made it but before the driver and his helper unloaded the bricks, drops of rain fell. It's just a shower, people said, and the crew retreated under cover to drink *tereré* until the storm passed.

It didn't pass for days. The truck load of bricks remained parked beneath the branches of the *paraiso* trees in the yard. The driver and his helper stayed at our house. His helper walked out the second day. The driver walked out after a week, leaving his loaded truck in our yard beneath the *paraiso* trees.

Meanwhile, as I waited to get the school kitchen construction project started, I made a wood form and produced some *lozas* from barbed-wire, sand, gravel and cement. I sold them at cost to those who wanted them. Most people who bought them didn't use them, I later learnt. They bought them to help me out, or so it seemed. Juan Enciso did construct a new latrine and used a concrete *loza*.

The teachers organized a soccer tournament to raise funds for the school and the community in general. Rain and mud did not dishearten the players and spectators. I taught the teachers and the students how to make doughnuts and we made about 100 on a *fogón*, most of which we sold. Fatima participated and then left for Durazno with relatives as soon as the school year ended.

***

Vilma announced her retirement from the school after twenty-three years there. She also announced that she had been communicating with Dr N_____ and she planned to become the director of the health centre. Some sauntered away upon hearing this news. Vilma had been accused of having students work for her without reward, of spending school funds on her own projects, and, just prior to her announcement, in this book-poor country, all the books I had collected from donors for the school library had vanished, as did a map of the community I had drawn in ink on Mylar and had given to the school.

I didn't know how the books and map flew, but I knew that I spent many days mapping the fifteen square kilometres of Guazú Cuá and drawing by night on an improvised drafting table. I'd used the copy of the original plat as a guide, improved upon it and changed the scale. I had passed many days visiting embassies in Asunción. Notwithstanding the losses, Vilma's retirement concerned few residents but many congratulated Nidia as the new director.

***

I met with Juan and Jorge beneath the shade of the *paraiso*s on a sweltering afternoon in early December of 1990. We discussed our plans to build a school kitchen while Jorge served *tereré*.

Juan and Jorge had already agreed to build a school kitchen with funds from my grant, but storms and the late arrival of cash delayed the project. I knew the men well enough to hire them for the school project. I'd worked with Jorge before and the three of us had become close as members of the performing arts group.

We decided to begin the following Monday morning.

"What time do you want to start?" I said.

"*Temprano*," Juan said.

"But what time is *temprano*," I said.

"*Tempranito*," Juan said.

"Give me a time so I know when you'll be here."

Juan shrugged and glanced at Jorge.

"We'll be here at 6:00," Jorge said.

"We'll get someone to haul the bricks to the school in a *carreta*," Juan said, "and stone for the foundation, too."

Jorge dumped the used *yerba* from the *guampa* and then splashed some water in to rinse it. We shook hands and Juan and Jorge went their respective ways. We'd meet again Monday morning to build a school kitchen, which would be situated beside the school and between the well and the garden. The teachers had started a garden project with the students.

We had already installed an NGO-donated hand pump in the well and we pumped water to an elevated tank. From the tank we would gravity feed water to a sink in the kitchen. Vegetables sprouted from the manure-rich soil of the garden. Students built a fence from branches and bamboo tied together with vines. As I learnt from my first garden at the health centre (not a trace of it left), without a fence pigs and chickens would enter and destroy the neatly planted rows.

As the sun broke the shadows on Monday, I waited beneath the *paraiso*s for Juan and Jorge.

My punctuality kept me waiting for Juan and Jorge to

begin the foundation for the school kitchen. I still had not tuned out my conditioning, or my misconception, to accept the concept of time prevalent in Guazú Cuá. An hour or so on one side or the other of a time to meet was common. A half a day wasn't unusual. Now I would be tested as well in my capacity as labourer-supervisor-grant administrator.

I sat some hours until Jorge arrived. He wore a tattered straw hat, a collage of a shirt with more patches than original material, thin brown pants and worn flip flops. In one hand he carried a bucket with a trowel, a roll of string, a plumb bob and a hammer. In the other he had a machete, a shovel and an iron bar.

Jorge greeted me, dropped his tools and took a seat. He rolled a cigarette, lit it and smoked while we waited for Juan. He showed up several minutes later with the same tools as Jorge plus a measuring tape and a length of clear surgical tubing which would be filled with water and used as a level. Juan, too, wore a straw hat. His old clothes had more life left than Jorge's, and Juan wore shoes.

Jorge gathered his tools and we walked a block to the school. I carried a five-gallon bucket, a pitcher, a *guampa*, a *bombilla* and *yerba mate*. My job as labourer-supervisor-grant administrator had begun. I would work under my employees as their labourer while I supervised construction and accounted for the grant.

***

At the school, a mound of sand, a bigger mound of red dirt, a pile of rocks and a stack of bricks waited to take the form of a school kitchen. Inside a classroom we had sacks of cement and lime, corrugated roofing material, doors and wood windows with frames, rafters, a sink, plumbing supplies, and the items needed to build a *fogón*.

We dropped our tools in the shaded corridor of the school. Jorge took his machete and started cutting branches for stakes from a nearby tree. Juan handed me a shovel and showed me where to dig a lime pit.

"Make it a metre by a metre and a half a metre deep," he said, kicking a rough border with his heel. I started to dig. Jorge whittled points on the stakes he had cut. He and Juan staked the area for the foundation squaring it by using "3-4-5" triangles and by measuring diagonals. When they finished Juan called me to stop, winked unobtrusively at Jorge, and told me to get water for *tereré*.

We sat on the grass in the shade beside the school and drank *tereré*. Juan and Jorge told me about dissent within the community because I'd hired them for the project and I would be the labourer-supervisor-grant administrator. People suspected that we were making money on the school kitchen. Only a couple of parents volunteered to help, such as Toribio, Antonio and a couple others, who hauled rocks, bricks and red dirt in ox-drawn carts.

That explained why residents who passed the school appeared not to notice us. I understood then a remark the former school director had made about my favourites. So began the project for the community: with most of the community against us.

Each morning Juan and Jorge met me beneath the *paraíso*s at Noemi's house and then we'd walk to work on the construction of the school kitchen. Not once as I dug, as I hauled rocks and bricks, as I pumped water and mixed mortar, did I imagine I'd see the kitchen walls rise and fall.

38

The first week, we dug a trench for the foundation. Juan and Jorge fitted rocks in the excavation and then we poured a soupy mortar over the rocks. We spent a week sweating in the shower room-like heat when the rocks reached a bit over ground level. We let the foundation set over the weekend.

I doffed my labourer's hat, washed my hands, and paid the builders and had them sign for the cash. I recorded the hours and the work accomplished in a project journal.

Monday morning I watched distant grey clouds building. We mixed a batch of mortar and began to set a double-wide layer of bricks on the stone foundation, tier after tier until we had a level surface on which to raise the walls. We smelt rain and saw downpours but only a refreshing drizzle fell on us.

As the wall rose, Juan and Jorge kept me busy supplying them with bricks and mud. I pumped water and mixed mortar, only stopping to drink *tereré*. The adhesive for the walls consisted of mostly red dirt with some cement and lime. I complained about the puny amount of cement we used, but was assured that it was sufficient with the red dirt. Each day grey and black clouds formed bringing the clean scent of precipitation and a strong breeze to revive the oppressive air.

Now the walls stood at the bottom height of the window frames. I loaded bricks on my bare arm to carry to the bricklayers. I lifted a brick and as I placed it on my arm I caught a black flash. A jet black, two-inch scorpion darted along the brick, tail raised, intending to strike. I dropped the brick and smashed the scorpion with my heel. I looked at Juan and Jorge looking at me. That was the last brick I handled without checking for scorpions.

Three weeks into the project, Juan and Jorge had built the walls to the height of the top of the door and window frames. At that level they'd next set a tier of rebar and concrete before adding more bricks. They used the rafters to support the walls in the interim. We cut and tied rebar to set the following day.

We left the school during a drizzle with a cool breeze that blew from nearby blackness. Just after dark, I sat drinking *mate* with Noemi as thunder rolled across Guazú Cuá. Flashes of lightning exposed dark houses. Wind bent branches and drops of rain splattered with a steady beat until later when it whipped in sheets.

Two walls fell almost completely during the storm; the other walls, partially. A mess of bricks and mortar surrounded what had been the school kitchen walls. I hadn't anticipated such an act of God in my proposal.

As we kicked among the debris, men with trowels joined us and they began to scrape the mortar from the fallen bricks to prepare to rebuild.

I saw the walls rise and fall. Juan, Jorge and two other men toppled the remnants of walls that hadn't fallen but had lost stability. Four adolescent boys scraped mortar from bricks. Another man pumped water to clean bricks and to use for fresh mortar. I helped clean and stack bricks.

"If I knew you were building such weak walls I would have done it myself," a man joked as he worked mortar with a hoe.

"If we knew that we would have let you do it," laughed Jorge.

"Leave that hoe and get a trowel then," Juan said as he slung red mud on the first layer of bricks. "Here's your chance to do it yourself."

Two girls arrived from the Nuñez house. "My mother sent empanadas and mandioca," one said, holding a cloth-covered basket. She placed it on a stack of bricks. "Help yourselves while they're warm. We'll come back for the basket later."

We ate and then started passing a few *guampas* of *tereré*.

Toribio passed a *guampa* to the man next to him. "Why don't Juan and Jorge and a couple of other good bricklayers build the walls and the rest of us can keep them supplied with mortar and bricks?" he said.

"I'm not sure I want to work with Juan and Jorge," a man said. "I thought they were experts. Look at the mess they made," he laughed.

"Let's get it done," Juan smiled.

After the walls were rebuilt, the work fell to me, to Juan and to Jorge again. We did have significant volunteer help from the brothers, Toribio and Clotilde, who felled two trees in the woods for *vigas*, or beams. They dragged the logs to the school by hitching oxen to a two-wheeled cart with a beam fixed above the axle. One end of a log they strapped to the beam and the other end dragged on the ground, pulled by a team of complacent, competent, muscular oxen. It seemed a long haul, but I'd seen teams of oxen pull four-wheel drive trucks buried to the axles. In two trips Toribo and Clotilde and their oxen delivered two logs to the school.

With ropes and brawn we lifted a log and fixed one end in the fork of a tree and the other sat on two posts set in the ground and tied together to form an X. Toribio, the wiry, smiling younger brother, barefoot with a tattered straw hat, his curly black hair protruding from the brim, in ragged jeans and a thin, long-sleeved shirt, hefted himself atop the log walking its length. He studied the log while Clotilde, the elder brother, a wide-brimmed hat covering his thinning grey hair, unhitched the oxen and walked them to pasture. He soon returned with a two-metre-long, two-handled cross cut saw with crocodile-like teeth across his shoulder. The wood handles held fast with wire and bolts.

Toribio, still barefoot and smiling, took one of the handles and Clotilde, on the ground and wearing shoes, held the other. Toribio illustrated his plan with gestures and the two sawyers began their cut, sweating, joking, and stopping for plenty of *tereré* breaks, during which they sharpened the teeth of their saw with files.

By the end of the following day, using only the two handled cross cut saw, a wedge and their eyesight, they'd squared both logs. The vigas weren't quite square, yet remarkably well-done, if not a bit rustic, considering they were cut by hand. Now we had the vigas we needed for the roof and verandah of the school kitchen.

On we laboured, simmering in the sun, to raise the kitchen at *Escuela Basica #778 Don Vicente Rolon*, Guazú Cuá, Escobar, except when heavy rains fell with lashing winds or when Juan needed time in his fields. We broke for the holidays as well.

Just before we quit for the Christmas season, a dirty white Mercedes Benz with four-wheel drive skidded to a stop in front of Antonio's store. The driver's door opened and the doctor emerged with a camera. He snapped several photographs of us and the school kitchen and leapt back into his vehicle, swung it about and left Guazú Cuá, his vehicle swerving and bouncing through ox cart ruts amidst spirals of dust. I never saw him again, but I later saw the photographs in his project binder at the GTZ office.

***

The holidays stopped our work. Christmas, New Year's Eve and New Year's Day, and the Day of the Three Kings, January 6, passed much like they had the year before. Most families built *pesebres*, Nativity scenes. Not far from the *pesebres* families kept a bucket of *clerico*. Christmas Eve, the people sauntered from house to house tasting *clerico* and respecting *pesebres*.

Christmas day saw grey plumes of smoke twisting up from cooking fires and beehive ovens. People sat beneath the shade of trees, the only relief from the blistering heat. Some shared *tereré* while others continued with the *clerico*.

Few worked at jobs or in the fields during these holidays. The New Year brought numerous parties under the glistening, starry sky. With no electricity for miles, the night sky performed a brilliant show of shooting stars, glimmering constellations and orbiting satellites. Voices singing to the strumming of guitars and shouts of drunkards radiated through the trees and the blackness of the night, save the glow of candles or cooking fires like ground-level stars. Fireworks spoilt the silence with the whistle of bottle rockets and the explosions of powerful firecrackers.

***

As my Peace Corps time would soon end, we planned to relocate to the United States. I had already started the tedious process of acquiring visas through the embassy in Asunción for Noemi and Augusto. We planned a church wedding for

March as well, at which all the people who helped me in Guazú Cuá would comprise the wedding party. Juan and Luisa Enciso agreed to be our *padrino and madrina*.

***

After The Day of the Three Kings, we returned to work on the school kitchen. We placed the *vigas* and secured them to the bricks with wire. We set hardwood rafters at right angles to the *vigas*. On the rafters we nailed sheets of corrugated roofing material with a row of ceramic tile and mortar along the crown of the roof.

Inside we pounded the earth with wood blocks and then covered the compacted area with chunks of broken bricks. This we covered with a layer of concrete, the base for a brick floor. I mixed sand, cement, and lime from our lime pit and hauled it by the bucket to Juan and Jorge who used the mortar to plaster the walls inside and out. Later we white-washed the plaster with a lime and water mix. I continued to supply Juan and Jorge with materials for the floor, to build a *fogón* and to install a sink. I pumped water into buckets, carried them near the lime pit, and there mixed mud. When I wasn't mixing I carried bricks. I dripped sweat. A downpour wouldn't have gotten me wetter.

I felt good about my work and my body felt strong, even though I'd lost about thirty of my previous 150 pounds since I'd been in Guazú Cuá. Work, walking and countless bouts of diarrhoea from amoebas dropped my weight. I thought I'd miss Guazú Cuá, but not the hours spent squatting in stinking outhouses voiding my bowels only to squat again after re-hydrating from my fluid loss.

I thought about how alienated I remained from what was normal for so many in the world. I knew life without electricity, plumbing, potable water and decent roads would soon end for me, but for the people of Guazú Cuá such life would continue indefinitely. The majority of the people in the world lived without basic services. I tolerated the inconveniences because they were temporary. Most people didn't have a choice.

We finished the school kitchen and Nidia organized a community party to inaugurate the new building. The kitchen and a few fundraisers would be my final projects as a Peace Corps volunteer in Guazú Cuá.

***

The inauguration party raised funds for the school while entertaining the community. We baked cakes in the school's new *fogón* to raffle. Volunteers prepared salads and *mandioca*. Others watched slabs of beef that hung from stakes drip fat, sizzling, into an open pit of coals. Wood smoke rose from the pit and the aroma of charred fat hung in the humid air. Sweating men moved and turned the spitting meat until it cooked through. Tethered, saddled horses grazed, their tails swishing at flies, along the school's fence line.

People lined up near the kitchen porch to buy plates of beef, salad and *mandioca*. Many piled salt beside their meat. A tub of cold well water cooled bottles of soda and beer. Families sat at tables and at student desks to tear into the tough beef and gnaw on bones before throwing them to skinny dogs. Juan, Desi and Silvio sat in the shade of the verandah sharing a beer and a flask while they tuned their guitars. Nidia stood and clapped her hands and called for attention. She stood on a chair and expressed her gratitude for the school pump and kitchen and for the support of the community, especially when the partially built walls of the kitchen fell to excessively strong winds during a storm. She thanked the community for their attendance and for their participation in fundraising activities such as this inauguration of the school kitchen. She stepped down.

The guitarists strummed and picked a Paraguayan polka, singing in Guaraní. On the grass, couples stepped to the music. Children with bare, dirty feet chased one another and climbed trees. A few women washed dishes in the new sink, and men in broad-brimmed cloth hats and wide woven belts wrapped between their jeans and shirts sat in the shade sharing glasses of beer. Dancing and drinking continued all afternoon, and then families drifted off on foot and on horseback. The inauguration proved successful despite the stifling heat of a February day.

As we left the school and passed Antonio and Vilma's house, Vilma called to us. We stopped and she told us that we'd have to take a course on marriage required by the Catholic Church before we could complete the fifth sacrament. She represented the Church in Guazú Cuá, so she would be our instructor. We'd meet at the *capilla* twice a week for three weeks.

I considered Vilma, the woman with whom I had been at odds almost since I'd arrived in Guazú Cuá two years prior. Perhaps I'd judged her too harshly, as if I were qualified to judge anyone. Yet the negative waves crashed against the walls of my skull. What could she teach us?

***

I also considered my own hypocrisy, in that, although I had been born and bred Catholic, I didn't include myself as a follower of the Church. But my Peace Corps service would soon end and we had decided to marry in a Catholic

Church before leaving Paraguay. This late decision created plenty of work for my sister, Fran, who acquired all the certificates—baptism, first communion, and confirmation—that the Catholic Church required before I could marry.

As it turned out there wasn't much to the classes. Vilma questioned us from a yellowed, mouldy booklet. Once we'd answered the questions about the sacrament of marriage, she awarded us a brittle, hand-written certificate onto which she pounded a rubber stamp authenticating the document and giving us permission to marry in the Catholic Church.

We set the date for March 3, 1991. We would marry at Santa Isabel, a leper colony several kilometres from Guazú Cuá. At Santa Isabel, while making wedding arrangements, I spoke with a Spanish priest who drilled me on the sacrament but was then inspired by my visit when I offered to have our performing arts group do a show for the residents of the colony after the wedding ceremony. He thought it would be good for the men and women afflicted with leprosy and exiled by the government to this remote site.

39

Our wedding day, March 3, 1991, arrived with a bright sun shining from a powder blue sky but with bearable heat. The night before, Rodolfo and Antonio dug a rectangular pit for a charcoal fire. Rodolfo had cut straight branches and stacked them near the pit. The men hacked the bark from the branches with long knives. They whittled points on each stake from which slabs of beef would hang over the fire pit for *asado de estaca*. *Don* Icho rode his horse into the yard leading a hefty young steer. He dismounted and tethered the Brahma to a post. Rodolfo helped him hobble the skittish grey bovid with lengths of braided strips of leather.

In the dim darkness before dawn, *Don* Icho, Rodolfo and Antonio forced the young animal to its knees and lashed it tightly to the post. One man held an aluminium basin ready as *Don* Icho make one slash across the squirming steer's neck with a sharpened blade. Blood gushed into the basin and slowed to a drip when the beast stopped trying to breathe. The men loosened the ties and raised the carcass by its hind legs so it hung from the post to drain the rest of its blood. The men carried the basin of blood to the kitchen where *Ña* Buena and other women waited to make the traditional blood sausage. The men exchanged the basin for another for the entrails. By the first light of day the men straddled the carcass, now on the ground, and skinned and butchered it. The internal organs were sent to the kitchen to be cleaned and cooked.

Inside the house, we prepared for the trip to the church at Santa Isabel. Noemi wore a simple elegant knee-length, white and pink dress that she had crafted. I wore beige pants and a white shirt with no tie. I didn't own one, anyway. Our wedding party gathered outside. It consisted of the sixteen members of our performing arts group, *Ña* Luisa, Juan's wife, Estela, who had come from Asunción and, finally, *Ña* Buena after she left the kitchen, as well as a few others.

We'd scheduled the wedding for 9:00 a.m. and we had about an hour's journey on foot and on horseback to the church. We carried some props and the guitars for our performance after the wedding. Other friends and family members stayed behind to cook the beef, stretch the hide, and to prepare soups and sausages with the entrails and the blood.

Our wedding party walked and rode beneath the morning's rising sun and covered the dusty seven kilometres to the church at *Colonia Santa Isabel* in about an hour. The short, grey-haired Spanish priest greeted us and we followed him into the basalt stone church. Apart from our group and the priest, two nuns accompanied us to assist with the ceremony.

The smiling, aging nuns shuffled about in their grey and white habits, lighting candles and arranging a Bible, a chalice and other materials on the altar.

The robed priest positioned us and our *madrina and padrino*. He then walked to the altar and the nuns sat, one on each side of the altar. The priest greeted the people and asked all to stand for an opening prayer.

After the prayer, with all sitting, he read a passage from the Book of Tobit in the Old Testament followed by a passage from John and then Matthew in the New Testament.

He spoke about Christian marriage and a nun joined him and held a copy of the text of "The Rite of Marriage." He asked all to stand and questioned Noemi and me about love and honour and faithfulness, reading awkwardly from the text. He then asked for our consent, saying, "What God has joined, men must not divide," and he blessed the rings—those that my father had given to me—that Juan and Luisa held and handed to us. We exchanged rings, the priest pronounced us man and wife, I kissed the bride and the wedding party clapped. After a closing prayer, the wedding party swarmed us with smiles, hugs and congratulations. They thanked the priest and the nuns and then filed out of the church.

We stayed so Jorge could take photographs of us with Juan, Luisa and the priest. By that time, outside, the rest of the group had prepared to perform for the permanent residents of the leper colony.

***

Beside the church an area paved with stone provided a space for outdoor gatherings. Buildings that housed the residents bordered this area, as well as buildings for staff and a medical facility. The residents sat in the shade of verandahs. The guitarists tuned up, Nidia, Zuni and Lita came out in costume and we began the show we'd done for many fund-raisers and fun for the last time.

Zuni and Lita, both barefoot, spun onto the paving stones, their long, billowing skirts and their braided hair lifting with their movements to a polka played from a cassette. Lita followed with a solo dance and then Nidia, dressed as a man, and Noemi sang their duet to the accompaniment of the guitars. Most of the group danced a polka, a waltz and a swing. We closed with a one-act play and a few tunes the guitarists played and sang. Nothing like it had been done before at *Colonia Santa Isabel*. Those who didn't have to go to *Colonia Santa Isabel* did not, so the residents had few visitors and no entertainment.

We laughed and talked our way back to Guazú Cuá on the dusty road, exhilarated from the event and heated under the mid-day sun.

At the house a small group waited. We had a few cases of inexpensive Argentine table wine to compliment the beef, salads and *mandioca*. Nidia and some of the girls had baked a cake with a score of pink ribbons hanging from it. Each single person held a ribbon and, on cue, pulled. The one who pulled a ribbon from the cake with a plastic ring on the other end would be the next to marry.

We drank wine with soda and danced on the grass in front of the house. We ate more of the abundant beef and the crowd drifted off before dark. We all enjoyed the time, the ceremony, the performance, and the party, yet we all sensed not just the onset of autumn, but also the change that the close of my Peace Corps service would bring to the community of which I'd become a member.

40

The final months of my volunteer service with Peace Corps Paraguay entailed many trips to Asunción and one to San Bernardino. I made those trips either alone, with Noemi, or with Noemi and Augusto, depending upon the task at hand. The visa process, for instance, took the three of us to various ministries to acquire documents or to have them authenticated. We purchased stamps, much like postage stamps, of varied denominations, sizes, and designs in blue, orange, purple, red or green. A ministry clerk charged with authenticating a document would wet the stamps on a sponge and affix them to the paper and then, with military precision, brandish a rubber stamp, slam it onto an ink pad and pound it onto a stamp—slam, pound, slam, pound—for each stamp. An official then signed over the stamps with loops, lines, curls and dots to represent an intricate yet illegible name. Often this process could not be completed the same day as one of the necessary public servants was inevitably out of the office.

I spent many hours sitting in a waiting room in the United States Embassy, listening to the counsel grant or deny visas and to assistants to the counsel explain the visa process to prospective travellers. As an American citizen, I entered the embassy grounds with more ease than noncitizens, who waited in a long line to gain access to the room and a chance to speak with an embassy employee, often only to be asked for more documentation. Although

I accessed the room without waiting in line, I still lingered with the others and listened for my name to be called. The counsel required that I have a business send a job offer to me through the embassy, and that someone sponsor us and prove his or her solvency by sending several years of tax returns and bank statements. I had to provide evidence that I could earn enough in the United States, something like 125% of the poverty level for a family of three. The sponsor would ensure that we did not use welfare. A former employer wrote a job offer for me and acted as our sponsor.

The Paraguayan government and the American Embassy obligated us to get permission to take a minor, Augusto, out of the country, meaning yet another document to be signed by both parents. Augusto's biological father, who'd abandoned the boy before birth and never contributed even a pinch of flour towards his support, refused to sign. This compelled us to hire a lawyer, who, in turn, dispatched a few young soldiers to the father's place of employment near Guazú Cuá. They took him into custody and escorted him to the lawyer's office in Paraguarí. There he had the choice of signing the document, giving Augusto permission to leave the country with us, or making a lump child support payment with interest to cover what he had failed to contribute for Augusto's welfare.

He didn't have the money, so the lawyer—legal or not—threatened to have him kept in custody until someone paid the sum for child support. He signed the document. We got a new stamp in Augusto's passport giving him permission to leave the country with us.

***

I returned Gulliver to *Don* Icho. I hadn't been riding him much and depended on my legs to get me here and there. One of the last times I rode Gulliver, we took the ox cart route from Paraguarí. We came to a flooded creek. I saw several ox carts on the other bank, loaded with merchandise and waiting for the water to go down to cross. Gulliver stepped into the creek. I though he wanted a drink, but he started across until he was swimming. I had two bags slung over the saddle. I put the bags on my shoulder and kneeled on the saddle as water washed over it and the current carried us downstream. I could see a group of people with the ox carts standing on the other bank watching. Gulliver kept swimming. A fence stretched across the creek at the high water line. Just before we hit it, Gulliver hit gravel and pulled us out of the creek. The ox cart drivers told me about the dangers of the creek, that no one crossed during high water, and that I must be insane for having done so. When I got home, I let Gulliver out to graze and only rode him a few more times.

***

In the intervening time, I attended several Peace Corps Paraguay meetings and workshops concerning Areguá-1's close of service. This process involved a medical screening and clearance, and writing descriptions of our particular jobs and achievements as volunteers—

embellished if one wished. We also wrote about our sites, making recommendations for future volunteers or advising Peace Corps Paraguay not to place another volunteer in a site, which is what I did.

I wrote that Peace Corps Paraguay should not send a second volunteer to Guazú Cuá. I explained that the population of the area continually decreased as the young—and sometimes entire families—left for the cities or other more promising places in the country. I also wrote about the remoteness of the site and that it fell far from the plan of having clusters of three volunteers within five kilometres of one another for support, safety and group projects. My cluster consisted of two women, five kilometres distant from one another, and me. My site sat at least twenty kilometres south of the other two, a distance difficult to travel directly because of the lack of transportation and a stretch of steep, rugged terrain. To arrive at their sites, I would trek fifteen kilometres to the bus stop and catch a bus another fifteen kilometres to Paraguarí. From there I'd take a second bus fifteen to twenty kilometres to their sites (they both lived on the same road along which travelled a bus a few times a day). So working with those in my cluster, or seeking help from them, or vice versa, in an emergency situation would not have been feasible. I had travelled to their sites only twice.

Peace Corps held one workshop in a hotel in San Bernardino on the shore of *Lago Ypacaraí*. There we talked about our return to the United States and reverse culture shock (which I'd already had a taste of during my emergency leave). We learnt about Peace Corps benefits such as graduate school fellowships and non-competitive status for federal jobs. We ended with a session on developing resumes to highlight our volunteer experience and accomplishments.

I made it to a couple of parties with my group to celebrate our close of service. I did so to see the remaining members of Areguá-1. Not half of the original group remained. I had alienated myself from the group during the previous two years since I had spent so much time in my site and so little time with other volunteers. I learnt most would return to the United States, some to work and others to study. At least one planned to remain for a third year. Two would leave as a married couple. Most seemed ready to leave Paraguay. I was ready to leave, yet I wanted to stay. But I didn't want to continue in my site and I'd already acquired visas for Noemi and Augusto.

I reflected on my Peace Corps experience and thought about the goals I had once explained to Jorge. I remembered the Peace Corps literature said volunteering would be "The toughest job you've ever loved." Certainly it proved challenging and despite the unending process of assimilation, in retrospect, I loved the job. I'd volunteer again, I thought, if my situation had been different.

Notwithstanding, the time arrived. My father still lived, so being with him again became a priority.

We had already purchased tickets to fly to Sacramento, California and planned to stay at my father's house in Grass Valley until I found suitable employment and while Noemi studied English and became acclimated to another culture. This, too, would allow my folks to know my family and for them to enjoy their grandson before my father succumbed to cancer.

We spent a night at Noemi's parents' house in Paraguarí and our last night in Asunción before we boarded our flight. As the jet ascended we looked below at the broad, murky, snaking Paraguay River and the red tiled roofs set against the divergent greens of the flora. I wondered about taking Noemi and Augusto to a land where advertisements dropped like carpet bombs and promoted the possession of things and the acquisition of more goods and services, often, if not usually, unnecessary. I thought of decent roads and indoor plumbing and electricity. What does one really need? As clouds obscured the view, I wondered when we'd return.

Vine Leaves Press

Enjoyed this book?
Go to *vineleavespress.com* to find more.

CPSIA information can be obtained
at www.ICGtesting.com
Printed in the USA
BVHW071932310321
603806BV00006B/624

9 781925 417661